C000140631

Higher

2

practice

for **AQA, Edexcel** and **OCR two-tier GCSE mathematics**

CAMBRIDGE
UNIVERSITY PRESS

The School Mathematics Project

Writing and editing for this edition John Ling, Paul Scruton, Susan Shilton, Heather West
SMP design and administration Melanie Bull, Pam Keetch, Nicky Lake, Cathy Syred, Ann White

The following people contributed to the original edition of SMP Interact for GCSE.

Benjamin Alldred	David Cassell	Spencer Instone	Susan Shilton
Juliette Baldwin	Ian Edney	Pamela Leon	Caroline Starkey
Simon Baxter	Stephen Feller	John Ling	Liz Stewart
Gill Beeney	Rosemary Flower	Carole Martin	Biff Vernon
Roger Beeney	John Gardiner	Lorna Mulhern	Jo Waddingham
Roger Bentote	Colin Goldsmith	Mary Pardoe	Nigel Webb
Sue Briggs	Bob Hartman	Paul Scruton	Heather West

CAMBRIDGE UNIVERSITY PRESS
Cambridge, New York, Melbourne, Madrid, Cape Town, Singapore, São Paulo, Delhi

Cambridge University Press
The Edinburgh Building, Cambridge CB2 8RU, UK

www.cambridge.org
Information on this title: www.cambridge.org/9780521689984

© The School Mathematics Project 2008

First published 2008

Printed in the United Kingdom at the University Press, Cambridge

A catalogue record for this publication is available from the British Library

ISBN 978-0-521-68998-4 paperback

Typesetting and technical illustrations by The School Mathematics Project
Cover design by Angela Ashton
Cover image by Jim Wehtje/Photodisc Green/Getty Images

Using this booklet

This booklet provides well graded exercises on topics in the Higher tier up to the level of GCSE grade A*. The exercises can be used for homework, consolidation work in class or revision. They follow the chapters and sections of the *Higher 2* students' book, so where that is the text used for teaching, the planning of homework or extra practice is easy.

Even when some other teaching text is used, this booklet's varied and thorough material is ideal for extra practice. The section headings – set out in the detailed contents list on the next few pages – clearly describe the GCSE topics covered and can be related to all boards' linear and major modular specifications by using the cross-references that can be downloaded as Excel files from **www.smpmaths.org.uk**

It is sometimes appropriate to have a single practice exercise that covers two sections within a *Higher 2* chapter. Such sections are bracketed together in this booklet's contents list.

Sections in *Higher 2* that do not have corresponding practice in this booklet are shown ghosted in the contents list.

The information on pages 46, 58 and 88 about the formula page supplied for the GCSE examination applies to specifications from all boards in England, Wales and Northern Ireland. Though correct at the date of publication of this booklet, it should be checked against later GCSE specifications.

Marked with a red page edge at intervals through the booklet are sections of mixed practice on previous work; these are in corresponding positions to the reviews in the students' book. A special feature of the mixed practice sections in this booklet is that they also contain questions on work from *Higher 1*, so that important skills and concepts can be 'kept alive' in the run up to the GCSE examination.

 Questions to be done without a calculator are marked with this symbol.

Questions marked with a star are more challenging.

Answers to this booklet are downloadable from **www.smpmaths.org.uk** in PDF format.

Contents

continues >

1 Graphing inequalities

You need squared paper and graph paper for section C.

B Regions with vertical and horizontal boundaries

1 Write an inequality for each shaded region.

(a)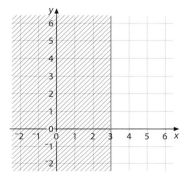

(b)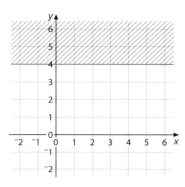

2 Write an inequality for each shaded region.

(a)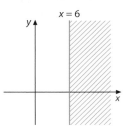

(b)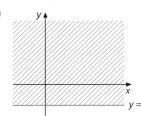

(c)

3 Draw a diagram for the region given by each inequality.

(a) $x \geq 1$ (b) $y \leq 3$ (c) $x \leq {}^-1$

4 Draw a pair of axes, each numbered from $^-2$ to 6.

(a) Draw the line with equation $y = 2$.

(b) Shade the region described by $y \geq 2$.

(c) On the same set of axes, draw the line with equation $x = 3$.

(d) Shade the region described by $x \geq 3$.

(e) (i) Show clearly the region described by both $y \geq 2$ and $x \geq 3$.

(ii) Give the coordinates of three points in this region.

5 A region is described by both the inequalities $x \leq 2$ and $y \geq 5$.
Give the coordinates of four points in this region.

6 Write down the four inequalities that define the shaded region.

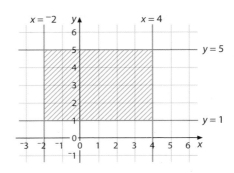

c Sloping boundary lines

1 Write an inequality for each shaded region.

(a)

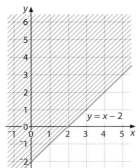

$y = x - 2$

(b)

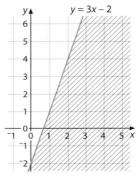

$y = 3x - 2$

(c)

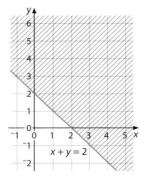

$x + y = 2$

2 Draw a pair of axes, each numbered from $^-2$ to 6.

   (a) Draw the line with equation $y = 2x - 1$.

   (b) Shade the region defined by $y \geq 2x - 1$.

   (c) Write down the coordinates of two points that satisfy the inequality $y \geq 2x - 1$.

3 Draw a diagram for the region given by each inequality. Number each axis from $^-2$ to 6.

   (a) $y \geq 3x - 2$ **(b)** $x + y \leq 5$ **(c)** $y \geq 1 - x$ **(d)** $3x + 5y \geq 15$

4 Find the equation of each line and write an inequality for each shaded region.

(a)

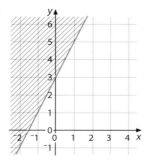

(b)

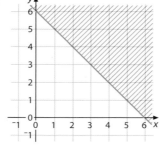

(c)

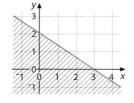

5 A teacher is taking some calculators and books to a meeting.
A calculator weighs 0.2 kg and a book weighs 0.5 kg.

(a) Write down an expression for the total weight in kg of x calculators and y books.

(b) The teacher cannot carry more than 10 kg.
Use this fact to write down an inequality in x and y.

(c) On graph paper show the region that represents this inequality.

(d) If the teacher wants to take the same number of books and calculators, how many of each can she take?

D Overlapping regions

1 Write down the three inequalities that define each shaded region.

(a)
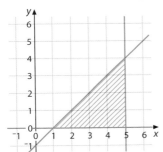

(b)
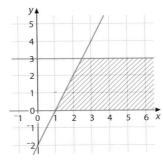

2 Draw a pair of axes, each numbered from 0 to 7.

(a) Show clearly the single region that is satisfied by all these inequalities.

$$x \geq 0 \qquad y \leq 5 \qquad y \leq x + 1$$

(b) Write down the coordinates of two points that satisfy all these inequalities.

3 Draw a pair of axes, each numbered from ⁻2 to 6.

(a) Show clearly the single region that is satisfied by all these inequalities.

$$x \leq 2 \qquad y \leq 5 \qquad x + y \geq 5$$

Label this region R.

(b) List all the points in R whose coordinates are integers.

4 Draw a pair of axes, each numbered from ⁻2 to 6.

(a) Show clearly the single region P that is satisfied by all these inequalities.

$$x + y \leq 3 \qquad y \geq 0 \qquad y \leq x$$

(b) Name the shape defined by region P.

5 Draw a pair of axes, each numbered from ⁻3 to 7.
Show clearly the single region that is satisfied by all these inequalities.

$$x \leq 6 \qquad y \geq {}^-2 \qquad y \leq 3x + 2 \qquad x + 3y \leq 6$$

6 Write down the three inequalities that define each shaded region.

(a)

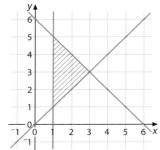

(b)

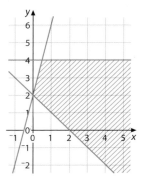

(c)

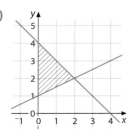

E Boundaries: included or not included

1 Write an inequality for each region.

(a)

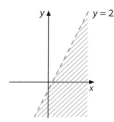

$y = 2x - 1$

(b)

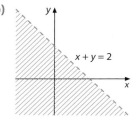

$x + y = 2$

2 Write down the two inequalities that define the shaded region.

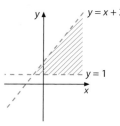

$y = x + 2$

$y = 1$

3 Draw a pair of axes, each numbered from $^-2$ to 6.

  (a) Show the region defined by these inequalities.

$$x > 0 \qquad y > x + 1 \qquad x + y \le 6$$

  (b) Which of these points lie in the region?

$$(0, 0) \quad (1, 3) \quad (1, 2) \quad (2, 1) \quad (2, 4) \quad (0, 4) \quad \left(2\tfrac{1}{2}, 3\tfrac{1}{2}\right) \quad (^-1, 0)$$

4 Draw a pair of axes, each numbered from $^-4$ to 12.

  (a) Show clearly the single region R that is satisfied by these inequalities.

$$^-2 < x \le 1 \qquad y \ge x - 2 \qquad 2x + y < 8$$

  (b) Name the shape defined by region R.

2 Enlargement and similarity

You need squared paper in section E.

A Enlargement and scale factor

1 This shape is to be enlarged with scale factor 2.5.
Make a sketch of the enlargement, showing its dimensions.

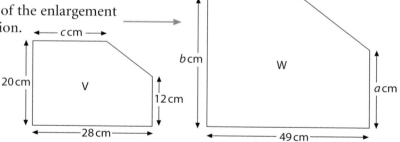

3.2 cm

2.2 cm

1.4 cm

3.0 cm   2.4 cm

2 Shape S is an enlargement of shape R.
(The shapes are not drawn accurately.)

(a) What is the scale factor
of the enlargement?

(b) Find a and b.

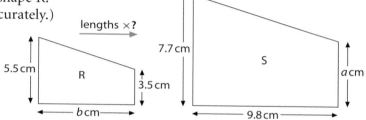

lengths ×?

5.5 cm   R   7.7 cm   S a cm

3.5 cm

b cm   9.8 cm

3 Shape W is an enlargement of shape V.

(a) Give the scale factor of the enlargement
as an improper fraction.

(b) Find a, b and c.

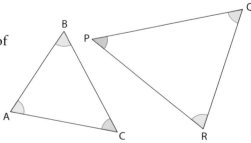

c cm   28 cm

20 cm   V b cm   W a cm

12 cm

28 cm   49 cm

B Similar triangles

1 These two triangles are similar.
Equal angles are marked the same way.
In triangle PQR, which side is the enlargement of

(a) AB   (b) AC   (c) CB

B   P   Q

A   C   R

2 (a) Explain why these two triangles must be similar.

(b) Find the scale factor that enlarges the smaller triangle to the larger one.

(c) Find the values of p and q.

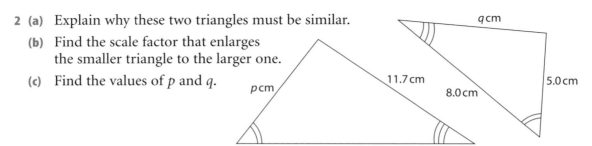

3 These triangles (not drawn accurately) are all similar to the blue triangle.

(a) What is the ratio $\dfrac{FG}{GH}$ in the blue triangle?

(b) Use this ratio to find the values shown by letters.

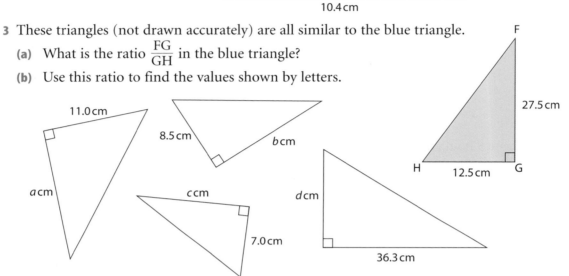

4 In the diagram MN is parallel to JK.

(a) Explain why triangles JKL and MNL are similar.

(b) Find length MN.

(c) Find length KL.

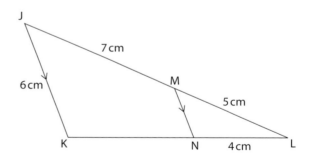

5 (a) Prove that triangles APY and BPX are similar.

(b) Find length AX.

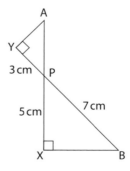

6 Calculate length AE to 1 d.p.

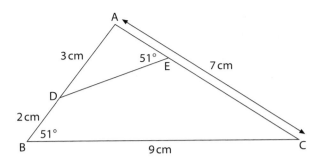

7 In the diagram PQ is parallel to RS.
PS = 12 cm.

    (a) Prove that triangles PQT and SRT
are similar.

    (b) Find length TS.

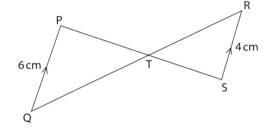

c Map scale

1 On maps with these scales, what actual distance in kilometres is represented
by a measurement of 1 cm?

    (a) 1 : 100 000   **(b)** 1 : 500 000   **(c)** 1 : 25 000   **(d)** 1 : 1 250 000

2 On a scale drawing, a door 2.1 m tall is drawn 42 mm tall.

    (a) What is the scale of the drawing, in the form 1 : n?

    (b) The door is 0.9 m wide. How wide is it on the drawing?

3 A map is drawn using a scale of 1 : 50 000.

    (a) What distance, in kilometres, does 1 cm on the map represent?

    (b) On the map, two towns are 6 cm apart.
What is the actual distance in kilometres between the towns?

    (c) On the map, a lake is 7.8 cm long.
What is the actual length of the lake in kilometres?

    (d) The distance between two churches is 4.9 km.
How far apart are the churches on the map?

4 Write each of these scales in the form 1 : n.

    (a) 1 cm to 10 km   **(b)** 1 cm to 500 m   **(c)** 2 cm to 8 km

5 Which of these scales do you think would be best to use for a road atlas?

    A 1 : 2500   **B** 1 : 25 000   **C** 1 : 250 000

1 A line is enlarged by a scale factor of 5.
   The resulting line is then enlarged by a scale factor of 3.
   What scale factor would take you in a single enlargement from
   the original line to the final line?

2 Shape P is enlarged by a scale factor of 4.5 to give shape Q.
   Shape Q is then enlarged by a scale factor of 1.4 to produce shape R.
   By what scale factor could shape P be enlarged to give shape R in one step?

3 A shape is enlarged with scale factor 3.
   Its image is then scaled down with scale factor 0.7.
   What scale factor would achieve the same effect in a single step?

4 A line is enlarged by a scale factor of 2.
   The resulting image is then enlarged, then its image is enlarged, and so on
   for several enlargements, always with a scale factor of 2.
   The final line is 32 times as long as the original line.
   How many enlargements have been carried out?

*5 Shape Z can be produced by enlarging shape X by a scale factor of 2.5.
   Shape Z can also be produced by enlarging shape Y by a scale factor of 2.
   How can shape X be obtained from shape Y?

E **Enlargement with a negative scale factor**

1 Copy this diagram.

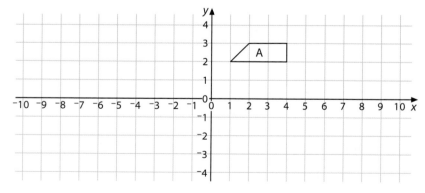

(a) Draw an enlargement of shape A using a scale factor of ⁻2 and
     (3, 1) as the centre of enlargement. Label this image B.

(b) Draw an enlargement of B with centre (0, 0) and scale factor ⁻1.
     Label the image C.

(c) What transformation maps A directly on to C? Describe it fully.

3 The reciprocal function

You need graph paper for section B.

A Review: reciprocals

1 What is the reciprocal of each of these?

(a) 7   (b) $\frac{1}{2}$ (c) $\frac{4}{3}$ (d) $\frac{3}{7}$ (e) $3\frac{1}{2}$

2 Find the reciprocal of each of these.

(a) 0.2   (b) 0.02   (c) ⁻0.00001   (d) 2.5   (e) 1.4

3 The reciprocal of 16 is 0.0625.
Write down the value of $\frac{1}{160}$.

4 Use the reciprocal key on a calculator to find the reciprocal of each of these as a decimal.

(a) 32   (b) 80   (c) 12.8   (d) ⁻625   (e) ⁻0.016

5 The reciprocal of x is 0.00390625.
Find the value of x.

6 (a) What is the value of $x^2 + \frac{1}{x}$ when $x = 4$?

(b) Use trial and improvement to find a solution to the equation $x^2 + \frac{1}{x} = 20$.
Give your answer correct to two decimal places.

B Graphs of reciprocal functions

1 (a) Copy and complete this table for $y = \frac{4}{x}$.
Give values correct to two decimal places where appropriate.

x	⁻4	⁻3	⁻2	⁻1	⁻0.5	⁻0.25	0.25	0.5	1	2	3	4
$y = \frac{4}{x}$	⁻1	⁻1.33				⁻16	16					

(b) (i) On graph paper, draw the graph of $y = \frac{4}{x}$ for values of x from ⁻4 to 4.

(ii) Use your graph to solve the equation $\frac{4}{x} = 13$, correct to 1 d.p.

2 (a) Find the value of $\frac{10}{x}$ when

(i) $x = 10000000$ (ii) $x = ⁻0.00000001$

(b) Sketch the graph of $y = \frac{10}{x}$ showing what happens for large and small positive and negative values of x.

4 Arc, sector and segment

A Arc and sector of a circle

1 For each of these, calculate

    (i) the length of the arc to the nearest 0.1 cm

    (ii) the area of the sector to the nearest 0.1 cm^2

(a)

(b)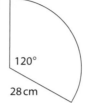

(c)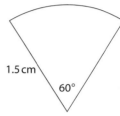

(d)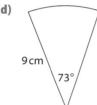

(e)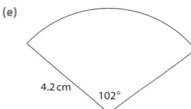

(f)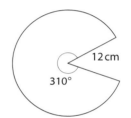

2 (a) Calculate, to the nearest 0.1 m^2, the area of a sector with radius 2.4 m and angle 62°.

    (b) Calculate, to the nearest 0.1 m, the length of an arc with radius 0.7 m and angle 265°.

3 Give each of these as an exact value, leaving π in your answer.

    (a) The area of this sector

    (b) The total perimeter of the sector

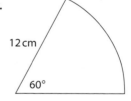

4 (a) Calculate the shaded area.

    (b) Calculate the total perimeter of the shaded shape.

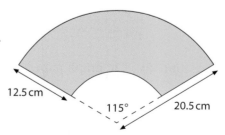

*5 Find the radius of a sector that has area 36 cm^2 and angle 75°.

*6 A sector with angle 130° has an arc of length 10 cm.
Calculate, to the nearest 0.1 cm, the radius of the arc.

B Segment of a circle

1 Calculate the area of each of the shaded segments.

(a)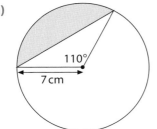

(b)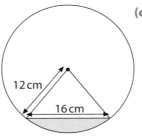

(c)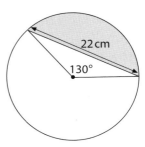

2 This earring design is based on an equilateral triangle.
The centres of the arcs are at the vertices.
Find the shaded area to the nearest 0.1 mm².

3 These segments are each more than half of a circle.
Calculate the area of each segment.

(a)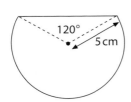

(b)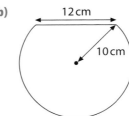

C Part of a cylinder

1 This solid is a sector of a cylinder.
Find, in terms of π, an exact value for its volume.

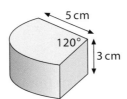

2 The cross-section of a solid is the sector of a circle
with radius 6 cm and angle 80°.
The height of the solid is 9 cm.

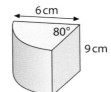

(a) Calculate the area of the sector of the circle.

(b) Calculate the length of the arc of the circle.

(c) Calculate the area of the curved surface of the solid.

(d) What is the total surface area of the solid?

5 Investigating and identifying graphs

B Identifying graphs

1 Here are nine equations and nine sketch graphs. Match them up.

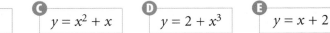

A $y = 2 - x^2$ **B** $y = 2x^3$ **C** $y = x^2 + x$ **D** $y = 2 + x^3$ **E** $y = x + 2$

F $y = 2 - 2x^3$ **G** $y = 2 - x$ **H** $y = \dfrac{2}{x}$ **I** $y = x^2 - 2$

P 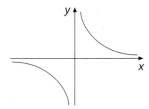 **Q** 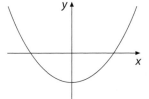 **R**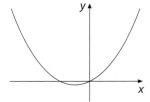

S 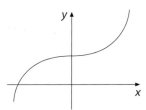 **T** 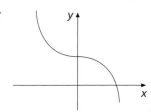 **U**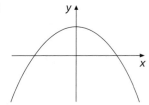

V 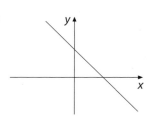 **W** 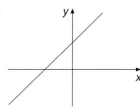 **X**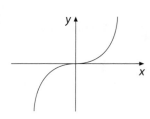

2 Select four equations from the following list to match the sketch graphs below.

A $y = \dfrac{-3}{x}$ **B** $y = x^2 + 3$ **C** $y = 3 - x^2$ **D** $y = 3x^3$

E $y = 3x + 1$ **F** $y = x^3 - 1$ **G** $y = (x + 1)^2$ **H** $y = 3 - x$

(a) 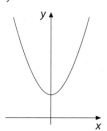 **(b)** 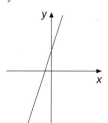 **(c)** 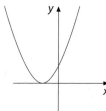 **(d)**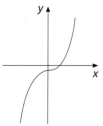

6 Combining probabilities

B Mutually exclusive events

1 One of these cards is picked at random.

| 14 | 15 | 16 | 17 | 18 | 19 | 20 | 21 | 22 | 23 |

Event J is 'The number picked is less than 20'.
Event K is 'The number picked is prime'.
Event L is 'The number picked is a multiple of 7'.
Event M is 'The number picked is a multiple of 4'.

(a) Are these pairs of events mutually exclusive?

    (i) J, K   **(ii)** J, L   **(iii)** K, L   **(iv)** K, M   **(v)** L, M

(b) What is the probability that the number picked is

    (i) either less than 20 or prime   **(ii)** either a multiple of 7 or a multiple of 4

    (iii) either prime or a multiple of 4   **(iv)** either even or a multiple of 7

2 A fairground game involves picking a card at random from a pack of playing cards.
A winning card is either an ace or a picture card (king, queen, jack).

Marie said:

> The probability of picking an ace is $\frac{4}{52}$.
> The probability of picking a picture card is $\frac{12}{52}$.
> So the probability of picking a winning card is $\frac{16}{52}$.

Is she right? Explain your answer.

3 This two-way table gives the numbers of types of books in a library.

	Fiction	Non-fiction
Hardback	80	120
Paperback	360	160

(a) A book is picked at random from the library.
Find the probability that it is

    (i) a hardback fiction book   **(ii)** a paperback non-fiction book

    (iii) a paperback book   **(iv)** either a hardback book or a fiction book

(b) **(i)** How many non-fiction books are there in the library?

    (ii) If a non-fiction book is picked at random, what is the probability
that it is a paperback book?

(c) If a paperback book is picked at random, what is the probability
that it is a non-fiction book?

1 In a game a player must roll an ordinary dice and spin the spinner.

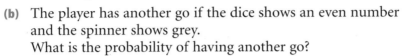

  (a) What is the probability that

   (i)   the spinner shows yellow

   (ii) the dice shows 4

   (iii) the spinner shows yellow and the dice shows 4

   (iv) the spinner shows blue and the dice shows 3

  (b) The player has another go if the dice shows an even number
      and the spinner shows grey.
      What is the probability of having another go?

2 Two ordinary dice are rolled. What is the probability that

  (a) both show 6   (b) neither shows 6

3 Two groups of people arrive at a hotel.
  The first group consists of 40 people of whom 15 can speak French.
  The second group consists of 50 people of whom 20 can speak French.

  A person is picked at random from each group.
  What is the probability that

  (a) both people can speak French   (b) neither person can speak French

D **Tree diagrams**

1 The probability of winning on a slot machine is $\frac{1}{15}$.
  Lisa has two goes on the slot machine.

  (a) Copy and complete the tree diagram
      for two goes on the slot machine,
      writing the probabilities on the branches.

  (b) Find the probability that Lisa wins on both goes.

  (c) Find the probability that Lisa wins on only one of her goes.

2 Kieran fits the battery and bulb into torches.
  The probability that the battery is faulty is 0.1.
  The probability that the bulb is faulty is 0.3.

  (a) Copy and complete the tree diagram
      for the torch.

  (b) The torch is faulty if either the battery
      or the bulb is faulty.
      What is the probability that the torch is faulty?

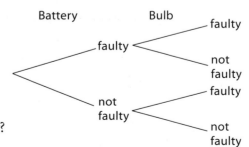

3 Two ordinary dice are rolled.
By drawing a tree diagram or otherwise, find the probability that
one dice shows 6 and the other does not show 6.

E Dependent events

1 A bag contains 4 red sweets and 5 green sweets.

(a) A sweet is taken out of the bag and not replaced. It is red.
What is the probability that the next sweet taken at random from the bag is

(i) red   (ii) green

Two sweets are taken at random from the bag without replacement.

(b) Copy and complete this tree diagram.

(c) Find the probability that

(i) both sweets are red

(ii) the two sweets are of different colours

(iii) at least one sweet is green

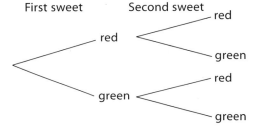

2 A box contains 4 black counters and 3 white counters.
Two counters are taken out at random without replacement.
Find the probability that

(a) both counters are black

(b) both counters are the same colour

(c) the counters are different colours

F Mixed questions

1 The probability of bus A being late is 0.2.
Independently the probability of bus B being late is 0.3.
What is the probability of at least one of the buses being late?

*2 This two-way table gives the numbers of types of
people in a family group.

Two people are picked at random from the
group (without replacement).
Find the probability that

	Male	Female
Adult	2	3
Child	5	6

(a) both people are adult males

(b) both people are adults

(c) one person is male and the other female

(d) at least one person is a female child

Mixed practice 1

You need squared paper.

1 Ann has a part-time job and is paid £5.85 for each hour she works.
In May she worked 61 hours. Estimate Ann's total pay for May.

2 Paula plants 15 seeds and 11 of them grow into plants.
What percentage of the seeds grow into plants?

3 Lee bought two bars of chocolate at 49p each, one chicken for £5.20,
1.5 kg of carrots at 28p per kg and some packets of crisps at 34p each.
He paid with a £10 note and received £1.02 change.
How many packets of crisps did he buy?

4 P and Q are similar shapes.
Find the values of a and b.

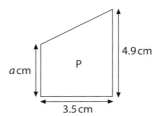

5 Solve these inequalities.

(a) $3x - 8 < 7$ (b) $10 - 2x \geq 1$ (c) $3(x - 2) < 4x + 1$

6 In a box there are some red, blue, yellow and green counters.
Jake is going to pick a counter from the box at random.
This table shows the probabilities of picking red, blue or yellow.

Colour	Red	Blue	Yellow	Green
Probability	0.3	0.15	0.2	

(a) Find the probability that Jake will pick a green counter.

(b) Find the probability that he will pick a counter that is not red.

7 Find the equation of the line parallel to $y = 3x - 5$ that goes through the point $(0, 2)$.

8 Express the area $9\,000\,000\,\text{cm}^2$ in m^2.

9 This bar chart shows the
times taken by a group of
people to complete a puzzle.

(a) How many people are
there in the group?

(b) Calculate an estimate
of the mean time taken.

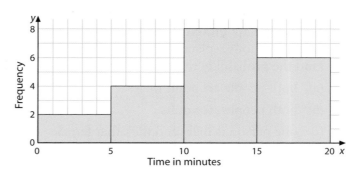

10 Both these spinners are spun.   A   B

(a) Copy and complete this tree diagram.

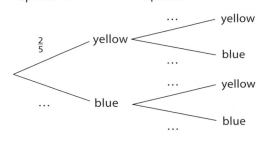

Spinner A   Spinner B

$\frac{2}{5}$ yellow

...   yellow

...   blue

...   blue

...   yellow

...   blue

(b) What is the probability that both spinners point to the same colour?

(c) What is the probability that the spinners point to different colours?

11 Simplify each of these.   **(a)** $\dfrac{4x + 6}{2}$ **(b)** $\dfrac{x^2 - 3x}{x}$

12 Find, as a fraction, the reciprocal of 1.4 .

13 (a) Match each graph with an equation on the right.

A 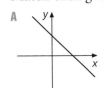   B   C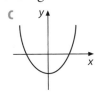

$y = x^2 + 1$ $y = x^2 - 1$

$y = 1 - x^2$ $y = x^3$

$y = \dfrac{1}{x}$ $x + y = 1$

$y = {}^-x$ $y = x^3 + 1$

D 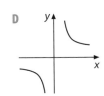   E   F

(b) Sketch a graph for each of the two unmatched equations.

14 Which of these scatter diagrams shows

(a) positive correlation   **(b)** negative correlation   **(c)** no correlation

A 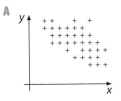   B 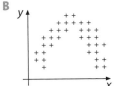   C 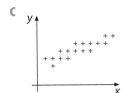   D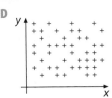

15 In this diagram QS is parallel to PT.

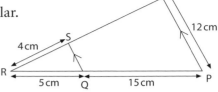

   (a) Explain why triangles PTR and QSR must be similar.

   (b) Calculate the length of QS.

   (c) Calculate the length of ST.

16 John estimates that the probability of his bus to work being late is 0.14.
Calculate the probability that his bus will be late two days in a row.

17 Expand and simplify $x(x + 3) - x(x - 5)$.

18 Copy this diagram.

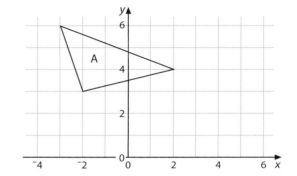

   (a) Draw an enlargement of triangle A with centre $(^-2, 0)$ and scale factor $\frac{1}{2}$. Label the image B.

   (b) Draw an enlargement of triangle A with centre $(0, 4)$ and scale factor $^-2$. Label the image C.

   (c) What is the scale factor of the enlargement from B to C?

19 Find the equation of the line that goes through the points $(0, ^-2)$ and $(2, 4)$.

20 Karen is playing bingo.
Balls numbered 1 to 100 are put in a box, picked one by one at random and called out.
Each player has a card with nine different numbers in the range 1 to 100.
What is the probability that the first two numbers called out are on Karen's card?

21 Draw a pair of axes, each numbered from $^-2$ to 9.
Show clearly the single region that is satisfied by all these inequalities.

     $x \leq 4$ $x + y > 2$ $y \leq 2x + 1$

Label this region R.

22 A map has a scale of $1 : 50\,000$.

   (a) The distance between Hangton and Petherton measures 8.5 cm on this map.
What is the distance between the two towns in kilometres?

   (b) The distance of Catsfield from Ninfield is 2.15 km at a bearing of 054°.
On this map, what is the distance and bearing of Catsfield from Ninfield?

23 Two representatives are to be chosen from a group of four girls and four boys.
The eight names are put into a box and two are drawn at random.
Calculate the probability that a boy and a girl will be chosen.

24 Find the gradient of the line with equation $3y - 2x = 6$.

25 (a) Find the equation of the line labelled a.

(b) Give the three inequalities satisfied by points in the region coloured red.

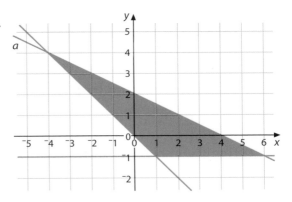

26 The diagram shows a sector of a circle of radius 12 mm. The angle of the sector is 40°.

(a) Calculate the perimeter of the sector.

(b) Calculate the area of the shaded segment.

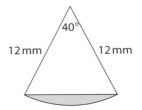

27 A manufacturer makes a locking mechanism containing a spring and a lever. The probability that the spring is faulty is 0.018 and the probability that the lever is faulty is 0.004. These probabilities are independent.

(a) In one day, the manufacturer makes 500 of these locking mechanisms. Estimate how many of them will have a faulty spring.

(b) What is the probability (to 3 s.f.) that in one locking mechanism,

    (i) neither spring nor lever is faulty

    (ii) at least one of these two components is faulty

28 Solve the equation $\frac{4}{9}x = \frac{1}{6}$.

29 Draw a scalene triangle.
Divide it into four congruent triangles that are similar to it.

30 A box contains 1 red counter, 2 blue counters and 3 green counters.
Sally takes a counter at random from the box but does not put it back.
Then she takes a second counter at random from those that are left.
Find the probability that both counters she takes are the same colour.

31 The diagram shows four squares 'nested' inside each other, with each square formed by joining the mid-points of the sides of the previous square.

(a) What fraction is the area of the fourth square of the largest square?

(b) The process continues until there are n squares. Write down an expression for the fraction that the nth square is of the largest square.

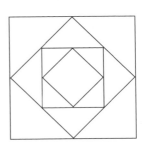

7 Direct and inverse proportion

B Quantities related by a rate
C Direct proportion

1 This graph shows the mass M kg of a type of oil against the volume V litres.

(a) What is the gradient of the graph?

(b) Write down the equation connecting M and V.

(c) Use your equation to find

(i) the mass of 8.5 litres of oil

(ii) the volume of 40.5 kg of oil

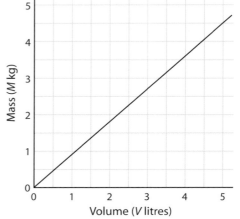

2 Are these statements true or false?

(a) The weight of a person is directly proportional to their height.

(b) The perimeter of a square is directly proportional to the length of the side.

(c) The mass of a piece of copper is directly proportional to its volume.

(d) The volume of a cube is directly proportional to the length of its edge.

3 In each of these tables, is Q directly proportional to P?
If so, give the equation connecting Q and P.

(a)

P	0	3	7	8
Q	0	12	28	32

(b)

P	1	2	3	4
Q	5	8	11	14

(c)

P	1	4	6	10
Q	6	24	36	60

(d)

P	0	1	3	6
Q	0	5	30	90

(e)

P	1	3	5	7
Q	8	10	12	14

(f)

P	2	6	14	22
Q	3	9	21	33

D Calculating with direct proportion
E $Q \propto P^2$, $Q \propto P^3$, ...

1 Y is directly proportional to X.
When $X = 8.0$, $Y = 4.8$.

(a) Find the equation connecting Y and X.

(b) Calculate 　　　　　(i) Y when $X = 3.5$ 　　　　　(ii) X when $Y = 10.8$

2 George is a decorator. The amount he charges for painting a ceiling is proportional to the area of the ceiling.

He charges £102.85 for painting a rectangular ceiling 6.8 m by 5.5 m.
How much does he charge for painting a ceiling 4.5 m by 3.6 m?

3 In this table, $Q \propto P$.

P	6.4	8.8		26.4
Q	11.2		21.7	

(a) Using $Q = kP$, find the value of k.

(b) Copy and complete the table.

4 The cost £C of a metre of cable is proportional to the square of the diameter d mm. When the diameter is 6 mm, the cost of a metre is £0.72.

(a) Find the equation connecting C and d in the form $C = kd^2$.

(b) Copy the table and fill in the missing values.

d	6	10	12	15	20
d^2	36				
C	0.72				

5 Q is directly proportional to P^2.
When $P = 4$, $Q = 8$.

(a) Sketch the graph of Q against P.

(b) Find the equation connecting Q and P.

(c) Use the equation to find the value of

(i) Q when $P = 8$

(ii) P when $Q = 72$

6 A truck is let go and runs down a slope.
The speed S m/s of the truck is directly proportional to the square root of the distance D m travelled.
When $D = 3.24$, $S = 1.2$.

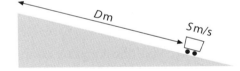

(a) Find the equation connecting S and D.

(b) Use the equation to calculate

(i) S when $D = 1.44$

(ii) D when $S = 2.0$

7 Given that Q is directly proportional to P^3 and that $Q = 500$ when $P = 5$, find

(a) the equation connecting Q and P

(b) the value of Q when $P = 4$

(c) the value of P when $Q = 32$

*8 If a gas is kept in a closed container, its pressure is directly proportional to its temperature, when the latter is measured in kelvin (K).

The rule for changing temperatures in °C to K is 'add 273'.
So, for example, 50 °C is equivalent to 323 K.

The pressure of the gas in a container at 15 °C is 1200 millibars.
What will the pressure be when the container is heated to 250 °C?
Give your answer to the nearest 10 millibars.

F Inverse proportion

1 When some gas is trapped and compressed, its volume, V litres, is inversely proportional to the pressure, P millibars, applied.

Here is a table of values of P and V.

P	1000	1200	1600	3000
V	4.8	4.0	3.0	1.6

(a) How can you tell from the table that V is inversely proportional to P?

(b) Sketch the graph of V against P.

(c) What is the equation connecting V and P?

(d) What is the value of (i) V when $P = 1500$ (ii) P when $V = 0.96$

2 If $Q \propto \dfrac{1}{P^2}$, and $Q = 10$ when $P = 5$, find

(a) the equation connecting Q and P (b) the value of Q when $P = 10$

3 $V \propto \dfrac{1}{U}$, and $V = 1.25$ when $U = 0.2$.

(a) Find the equation connecting V and U.

(b) Find the value of (i) V when $U = 10$ (ii) U when $V = 0.04$

4 y is inversely proportional to x.
When $x = 9$, $y = 4$.
Find the value of y when $x = 15$.

5 In this table, $Y \propto \dfrac{1}{\sqrt{X}}$.

X	16	25		100
Y	10		20	

(a) Find the equation connecting Y and X.

(b) Copy and complete the table.

G Identifying a proportional relationship

1 This table gives values of two quantities L and M.

L	2	4	6	8	10
M	2	16	54	128	250

(a) What is M multiplied by when L is multiplied by 2?

(b) What is M multiplied by when L is multiplied by 3?

(c) What is M proportional to?

(d) Find the equation connecting M and L.

2 In each of the tables below, Q is either proportional to P, P^2 or P^3, or inversely proportional to P, P^2 or P^3.

Find the type of proportionality for each table and the equation connecting Q and P.

(a)

P	1	2	3	6
Q	12	6	4	2

(b)

P	2	5	6	10
Q	12	75	108	300

(c)

P	1	2	4	8
Q	256	32	4	0.5

8 Quadratic expressions and equations 1

> **B Multiplying out expressions such as $(2n - 3)(4n + 5)$**
> **C Factorising quadratic expressions**

1 Multiply out the brackets in each of these and simplify the result.

   (a) $(2n + 7)(n + 3)$ (b) $(n + 4)(3n + 4)$ (c) $(3n - 2)(n + 1)$

   (d) $(4n - 5)(n + 1)$ (e) $(5n - 1)(n - 5)$ (f) $(6n - 5)(n - 3)$

2 Multiply out the brackets in each of these and simplify the result.

   (a) $(5x + 4)(3x + 2)$ (b) $(4x + 1)(2x + 5)$ (c) $(3x + 1)^2$

   (d) $(2x + 3)^2$ (e) $(7x - 3)(2x + 3)$ (f) $(2x + 5)(4x - 5)$

   (g) $(3x + 1)(2x - 3)$ (h) $(4x + 3)(2x - 5)$ (i) $(2x - 5)(5x - 4)$

   (j) $(4x + 1)(4x - 1)$ (k) $(6x - 5)(2x - 1)$ (l) $(5x - 2)^2$

   (m) $(3x + 2)(3x - 2)$ (n) $(4x + 3)(2 - x)$ (o) $(2 + x)(3 - 5x)$

   (p) $(5 - 2x)^2$ (q) $(3 - 2x)(3 + 2x)$ (r) $(1 - 2x)(5 - 4x)$

3 Show that the difference in area between the large and small rectangles is $2x^2 + 14x + 5$ square units.

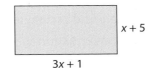

4 Find pairs from the loop that multiply to give

   (a) $2x^2 + 5x + 2$ (b) $3x^2 + 5x - 2$

   (c) $3x^2 - 5x - 2$ (d) $2x^2 - 3x - 2$

   (e) $6x^2 + 13x + 2$ (f) $6x^2 + 5x + 1$

   (g) $4x^2 - 1$ (h) $4x^2 - 9x + 2$

   (i) $9x^2 - 1$ (j) $6x^2 - 5x + 1$

Loop: $(3x - 1)$ $(x + 2)$ $(x - 2)$ $(3x + 1)$ $(2x + 1)$ $(2x - 1)$ $(4x - 1)$ $(6x + 1)$

5 Factorise these.

   (a) $2x^2 + 11x + 5$ (b) $3x^2 + 8x + 5$ (c) $3x^2 + 20x - 7$

   (d) $7x^2 - 15x + 2$ (e) $5x^2 - 17x + 6$ (f) $2x^2 + 9x + 10$

   (g) $2x^2 + 17x - 9$ (h) $3x^2 - 19x + 6$ (i) $11x^2 - 46x + 8$

6 Factorise these.

   (a) $4x^2 + 7x + 3$ (b) $4x^2 + 13x + 3$ (c) $6x^2 - 13x - 5$

   (d) $4n^2 + 3n - 1$ (e) $25n^2 - 1$ (f) $8n^2 + 6n + 1$

   (g) $6a^2 - 5a - 6$ (h) $9a^2 - 16$ (i) $4a^2 - 4a + 1$

   (j) $9y^2 - 3y - 30$ (k) $35y^2 - 125y - 60$ (l) $50y^2 + 60y + 18$

7 The nth term of a sequence is $9n^2 + 12n + 4$.

(a) Work out the first three terms of the sequence.

(b) Show that every term in the sequence must be a square number.

8 Show that, when n is an integer, the value of $4n^2 + 8n + 3$ can always be written as the product of two consecutive odd numbers.

D Solving equations by factorising

1 Solve these by factorising.

(a) $2x^2 - 15x + 7 = 0$ (b) $3x^2 - 5x - 2 = 0$ (c) $2x^2 + 9x + 9 = 0$

(d) $5x^2 - 7x + 2 = 0$ (e) $16x^2 - 1 = 0$ (f) $3x^2 + 12x = 0$

(g) $4x^2 + 7x + 3 = 0$ (h) $4x^2 + 28x + 49 = 0$ (i) $2x^2 - 10x + 12 = 0$

(j) $12x^2 + 8x - 15 = 0$ (k) $9x^2 - 24x = 0$ (l) $48x^2 - 80x + 12 = 0$

2 Solve these quadratic equations, rearranging them first.

(a) $2x^2 = x + 10$ (b) $5t^2 + 9t = 6 - 20t$ (c) $6y^2 = 11y - 3$

(d) $12c^2 + 1 = 32c - 20$ (e) $10r = 25 - 8r^2$ (f) $4 - 10a = 10a - 9a^2$

3 The triangle and rectangle both have the same area.

(a) Form an equation in x and show that it simplifies to $2x^2 - 13x + 6 = 0$.

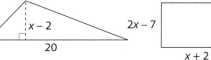

(b) Solve the equation and find the area of each shape.

E Quadratic expressions that use two letters

1 Expand and simplify these where possible.

(a) $(x + 5)(2y + 1)$ (b) $(2x + y)(x - 3y)$ (c) $(5y - 2)(2x + y)$

(d) $(3x - 2y)(x - y)$ (e) $(5x + y)^2$ (f) $(y - 5)(x - 3y)$

(g) $(4a - b)^2$ (h) $(5b - 2a)^2$ (i) $(3a - 2)(5a + 4b)$

(j) $(m - 4n)(m + 4n)$ (k) $(2m + 3n)(2m - 3n)$ (l) $(5m - 2n)(5m + 2n)$

2 Factorise these.

(a) $p^2 - q^2$ (b) $100y^2 - z^2$ (c) $49a^2 - 4b^2$ (d) $x^2 - 81y^2$

3 Factorise these.

(a) $10p^2 + 7pq + q^2$ (b) $6r^2 - 7rs - 3s^2$ (c) $8c^2 - 11cd - 10d^2$

4 Factorise these.

(a) $pq + p + q + 1$ (b) $2x^2 - 2xy + 3x - 3y$ (c) $a^2 + ab - a - b$

9 Dimension of an expression

A Length, area and volume

1 In the following expressions x, y and z represent lengths.
For each expression state whether it could represent a length, an area or a volume.

(a) $x + y$ (b) xy (c) xyz (d) y^2 (e) x^2z

2 One of the expressions below gives the approximate area of a regular hexagon with an edge length of s. Which one is it?

$3.142s$ $6.284s^3$ $2.598s^2$ $1.254s^6$

3 One of these expressions gives the volume of this solid.
Which one is it?

πabc $\pi a^2 b^2$

$\pi(a + b)c$ $\pi(a^2 + b)$

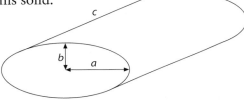

4 The letters f, g and h represent lengths.
2, 3 and 5 are pure numbers, which have no dimension.
For each expression, say whether it could represent a length, an area, a volume or a pure number, or is dimensionally inconsistent.

(a) $\dfrac{g + h}{2}$ (b) $\dfrac{3f^2}{h}$ (c) $fg + 5gh$ (d) $3f^3 - h^3$ (e) $(5f - g)^2$

(f) $\dfrac{3(g^2 + h^2)}{g + h}$ (g) $\dfrac{g^3}{2h} - \dfrac{h}{5}$ (h) $\dfrac{fg + gh}{3f}$ (i) $\dfrac{5fg}{h^2}$ (j) $f(3g + h)$

5 The letters p and q represent lengths.
Say whether each expression could represent a length, an area, a volume or a pure number, or is dimensionally inconsistent.

(a) $\sqrt{p^2 + r^2}$ (b) $\dfrac{\pi\sqrt{pq}}{r}$ (c) $\dfrac{\pi r^3}{\sqrt{p^2 - q^2}}$ (d) $\sqrt{p^2 - r}$ (e) $\left(\sqrt{pq}\right)^3$

6 The letters r, s and t represent lengths.

(a) What value must n have if $\dfrac{r^n}{s - t}$ represents an area?

(b) What value must n have if $\dfrac{r^2 s^2}{t^n}$ represents a length?

(c) What value must n have if $\pi r^n \sqrt{p^2 + q^2}$ represents a volume?

(d) What value must n have if $\dfrac{r^4}{s^n} + t^n$ is to be dimensionally consistent?

10 Sequences

A Linear sequences
B Quadratic sequences

1 Copy and complete each linear sequence, to show the first five terms.

(a) 4, 10, _____ , 22, _____ , ...

(b) 42, _____ , 34, _____ , 26, ...

2 (a) Find an expression for the nth term of each of these linear sequences.

(i) 7, 12, 17, 22, 27, ...

(ii) 2, 8, 14, 20, 26, ...

(iii) ⁻4, ⁻1, 2, 5, 8, ...

(iv) 37, 34, 31, 28, 25, ...

(b) Find the 10th term of each sequence in part (a).

3 A linear sequence begins 5, 9, 13, 17, ...

(a) Find an expression for the nth term of the sequence.

(b) What is the 20th term of the sequence?

(c) Show that 120 cannot be a term in this sequence.

(d) Which term in this sequence is 185?

4 (a) Find an expression for the nth term of each of these quadratic sequences.

(i) 6, 9, 14, 21, 30, ...

(ii) 3, 12, 27, 48, 75, ...

(b) Find the 100th term of each sequence in part (a).

(c) Show that 600 is not in either of the sequences in part (a).

5 A quadratic sequence begins 5, 9, 15, 23, ...

(a) Find the next two terms of this sequence.

(b) Find an expression for the nth term of this sequence.

(c) What is the 20th term in this sequence?

6 A sequence begins 1, 2, 6, 15, 31, ...
Could this be a quadratic sequence?
Explain how you decided.

C Mixed sequences

1 Write down the first three terms of the sequence whose nth term is

(a) $4n + 3$

(b) $n^4 + 2$

(c) $2n^2 + 4$

(d) $\dfrac{24}{n}$

(e) $\dfrac{n}{n + 3}$

(f) $n^3 - n^2$

2 Find the *n*th term of each of these sequences.

(a) 11, 14, 19, 26, 35, …

(b) 2, 6, 12, 20, 30, …

(c) 2, 6, 10, 14, 18, …

(d) 1, 8, 27, 64, 125, …

(e) $\frac{2}{3}, \frac{3}{7}, \frac{4}{11}, \frac{5}{15}, \frac{6}{19}, \ldots$

(f) 20, 17, 14, 11, 8, …

(g) 2, 8, 16, 26, 38, …

(h) $\frac{1}{4}, \frac{2}{9}, \frac{3}{16}, \frac{4}{25}, \frac{5}{36}, \ldots$

D Analysing spatial patterns

1 A sequence of matchstick patterns begins like this.

Pattern 1   Pattern 2   Pattern 3

Find an expression for the number of matches in pattern *n*.
Explain how you found this expression and how you know it is correct.

2 Points are marked on a circle, and chords are drawn joining each point to **every other** point.

| 1 point | 2 points | 3 points | 4 points |
| 0 chords | 1 chord | 3 chords | 6 chords |

(a) Find an expression for the number of chords if there are *n* points on the circle. Explain how you found this expression and how you know it is correct.

(b) How many chords can be drawn if there are 20 points on the circle?

(c) If 66 chords can be drawn, how many points are there on the circle?

3 Here are two sequences of matchstick patterns.

A
   …

B
  …

(a) For each sequence, find an expression for the number of matches in the *n*th pattern. Explain how you found each expression and how you know it is correct.

(b) Petra has 16 380 matches.

(i) Which pattern in sequence A could she make?

(ii) Which pattern in sequence B could she make?

11 Enlargement, area and volume

A Enlargement and area

1 Neil has an old photograph that is 40 mm by 60 mm.
He uses his scanner and computer to make an enlargement 120 mm by 180 mm.

   (a) What is the scale factor of the enlargement?

   (b) By what factor has the area increased?

2 A semicircle has an area of $18.0\,cm^2$.

   Work out   (i) the area factor   and   (ii) the area of the enlarged shape
   when the semicircle is enlarged with a scale factor of

   (a) 5   (b) 2.2   (c) $\frac{5}{3}$ (d) 0.8

3 Sphere Y has twice the diameter of sphere X.
The surface area of sphere X is $7.5\,cm^2$.
What is the surface area of sphere Y?

4 A garden designer draws a scale plan of a garden.
The area of the real lawn is 2500 times the area representing it on the plan.
What is the scale of the plan, in the form $1:n$?

5 What scale factor corresponds to each of these area factors?

   (a) 4   (b) 1.69   (c) $2\frac{1}{4}$ (d) 81

6 A sailmaker wants to make a sail that has 6 times the area of a sail on a model boat.
Which of these scale factors should he use?

   | 3 | 6 | $\sqrt{6}$ | $\sqrt{5}$ | 1.5 |

7 Two mirrors are of a similar shape.
The area of glass in the larger one is double that in the smaller.
The width of the larger mirror is 30 cm.
What is the width of the smaller mirror?

8 A jewellery designer makes a large scale drawing of a pendant she is going to make.
The real pendant will have an area of $9\,cm^2$.
The drawing has an area of $144\,cm^2$.

   (a) What is the scale factor of the enlargement from the actual pendant to the drawing?

   (b) What is the scale factor from the drawing to the pendant?

9 What is the area factor of a scaling down with a scale factor of

   (a) $\frac{1}{5}$ (b) 0.4   (c) $\frac{2}{3}$ (d) 0.9   (e) 0.01

10 The Cyclone wind turbine has blades with a total surface area of 350 m² facing the wind.

   (a) The manufacturers produce another wind turbine called the Tornado,
      which is the Cyclone design scaled down with a scale factor of 0.75.
      What is the wind-facing area of the Tornado's blades?

   (b) The Zephyr is also a scaled-down version of the Cyclone.
      Its blades have a wind-facing area of 126 m².
      What scale factor is used to scale the Cyclone down to the Zephyr?

11 Two circles have areas in the ratio 16:9. What is the ratio of their diameters?

B **Enlargement and volume**

1 Cuboids V and W are similar.
The scale factor from V to W is 1.1.

   (a) What is the volume factor from V to W?

   (b) Given that the volume of V is 2000 cm³,
      what is the volume of W?

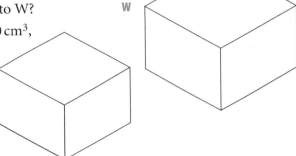

2 Find the volume factor for an enlargement whose scale factor is

   (a) 3   **(b)** 6   **(c)** 15   **(d)** 0.9   **(e)** 0.2

3 A working model of a steam engine is made to '$\frac{1}{8}$ scale'.
The water tank in the model holds 6 litres of water.
How much does the full-size steam engine's water tank hold?

4 These two glasses are similar.
The smaller glass holds 108 cm³ of liquid.
What does the larger one hold?

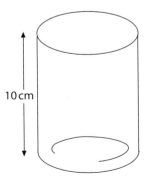

10 cm

6 cm

5 Find the scale factor of an enlargement whose volume factor is

   (a) 125   **(b)** 729   **(c)** 216 000   **(d)** 27 000   **(e)** 0.064

6 An enlarged version of a statue weighs 64 times as much as the original and is made of the same stone. Find the scale factor of the enlargement.

7 These two bread tins are the same shape.
One is for a 1 pound loaf, the other for a 2 pound loaf.
Which of these scale factors enlarges
the smaller one to the larger?

$1\frac{1}{2}$ $\sqrt[3]{1}$ $\sqrt{2}$ $1\frac{1}{3}$ $\sqrt[3]{2}$

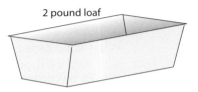

1 pound loaf

2 pound loaf

8 This carton contains cottage cheese.
The manufacturers want a larger, similar-shaped carton that they can label '50% extra'.

  (a) What is the volume factor for this enlargement?

  (b) Find the scale factor needed, explaining your method.

9 Two spheres have volumes in the ratio $27:8$. What is the ratio of their diameters?

10 F and G are similar objects, made from the same material.
F's surface area is 6.25 times that of G.
The mass of G is 100 g. Find the mass of F.

*11 Three similar objects have surface areas in the ratio $4:25:49$.
Give the ratio of their volumes.

c Mixed units

1 An architect draws a building to a scale of $1:50$.
A door is drawn 42 mm high and 18 mm wide.
How big in metres will the door be in real life?

2 A plan of a room is drawn to a scale of $1:20$.
A large rug in the room has an area of $7\,m^2$.
What area in cm^2 will this be represented by on the plan?

3 A road contractor is working from a plan on a scale of $1:500$.
A stretch of road to be resurfaced covers $32.5\,cm^2$ on the plan.
What is this area in m^2 on the real road?

4 A model for a sculpture is made on a scale of $1:20$.
The model has a volume of $8000\,cm^3$.
What will the volume of the finished sculpture be, in m^3?

5 A model for a statue has a volume of $8\,cm^3$.
The full-size statue has a volume of $1\,m^3$.
What is the scale of the model, as a ratio?

12 Algebraic fractions and equations 1

A Simplifying by cancelling
B Factorising then simplifying

1 Simplify each of these.

(a) $\dfrac{8x^2}{4x}$ (b) $\dfrac{5y}{10xy}$ (c) $\dfrac{12y^2}{8y}$ (d) $\dfrac{a^2b^3}{ab^{10}}$ (e) $\dfrac{6a^4b^3}{15a^9b}$

2 Simplify each of these.

(a) $\dfrac{9(x+5)}{x+5}$ (b) $\dfrac{5(x-2)^2}{x-2}$ (c) $\dfrac{x-3}{(x-3)^2}$ (d) $\dfrac{16(x+1)^3}{12(x+1)}$

(e) $\dfrac{3(y+4)}{(y+4)(y+1)}$ (f) $\dfrac{9y(y-10)}{(y-10)^2}$ (g) $\dfrac{(4y+3)^2}{(y+3)(4y+3)}$ (h) $\dfrac{7y(y-1)^2}{y(y-1)}$

3 Simplify each of these by factorising and cancelling.

(a) $\dfrac{3x+9}{x+3}$ (b) $\dfrac{10x-20}{x-2}$ (c) $\dfrac{5x+10}{15x+30}$ (d) $\dfrac{4x-6}{10x-15}$

4 Simplify each of these by factorising and cancelling.

(a) $\dfrac{x^2+3x+2}{x+2}$ (b) $\dfrac{x^2+8x}{x^2+16x+64}$ (c) $\dfrac{3x^2+11x+10}{3x+5}$ (d) $\dfrac{3x^2+9x}{2x^2+13x+21}$

(e) $\dfrac{x^2+5x+6}{x^2+4x+4}$ (f) $\dfrac{3x^2+14x+8}{x^2+9x+20}$ (g) $\dfrac{x^2-5x-14}{x-7}$ (h) $\dfrac{5x-25}{x^2-3x-10}$

(i) $\dfrac{x^2+x-12}{x-3}$ (j) $\dfrac{2x+2}{x^2-3x-4}$ (k) $\dfrac{12x-6}{4x^2-4x+1}$ (l) $\dfrac{x^2-9}{x+3}$

(m) $\dfrac{6x^2+x-2}{3x+2}$ (n) $\dfrac{9x^2-16}{6x+8}$ (o) $\dfrac{x^2+3x-10}{x^2+8x+15}$ (p) $\dfrac{5x^2-125}{x^2+2x-35}$

(q) $\dfrac{4x^2+3x-7}{12x^2-13x+1}$ (r) $\dfrac{4x^2-4x-3}{10x^2+3x-1}$ (s) $\dfrac{16x^2-100}{6x^2+11x-10}$ (t) $\dfrac{6x^2+7x-10}{18x^2-27x+10}$

5 (a) (i) Work out the value of $\dfrac{x^2+11x+10}{2x+20}$ when $x=2$.

    (ii) Work out the value of $\dfrac{x^2+11x+10}{2x+20}$ when $x=7$.

(b) Show that, if x is an odd number, $\dfrac{x^2+11x+10}{2x+20}$ is always an integer.

6 Simplify each of these by factorising and cancelling.

(a) $\dfrac{a^2-b^2}{a+b}$ (b) $\dfrac{15x+3y}{25x^2-y^2}$ (c) $\dfrac{m^2-9n^2}{m^2+3mn}$ (d) $\dfrac{3pq+6q^2}{3p^2-12q^2}$

c Multiplying and dividing

1 Simplify each of these.

(a) $\dfrac{8}{p} \times \dfrac{p}{4}$ (b) $\dfrac{p}{12} \times \dfrac{18}{p}$ (c) $\dfrac{p}{8} \times \dfrac{4p}{3}$ (d) $\dfrac{4p}{5} \times \dfrac{2}{3p}$

(e) $\dfrac{15}{4c} \times \dfrac{2}{3d}$ (f) $\dfrac{3ab}{4} \times \dfrac{2a}{b}$ (g) $\dfrac{4c}{d^2} \times \dfrac{cd}{2}$ (h) $\dfrac{3a^2}{4b} \times \dfrac{2b^2}{a}$

(i) $\dfrac{2a+1}{a} \times 3a$ (j) $2r\pi \times \dfrac{5}{6rs}$ (k) $\dfrac{a}{2a-4} \times a-2$ (l) $4a^2b \times \dfrac{3}{2ab^3}$

2 Simplify each of these.

(a) $\dfrac{8}{p} \div \dfrac{p}{4}$ (b) $\dfrac{p}{8} \div \dfrac{p}{6}$ (c) $\dfrac{2n^2}{3} \div \dfrac{1}{n}$ (d) $\dfrac{3ab^2}{4} \div \dfrac{a}{b}$

(e) $\dfrac{5ab}{3} \div \dfrac{10b}{a}$ (f) $\dfrac{4a^2bc}{3} \div \dfrac{2ab^2c}{5}$ (g) $\dfrac{1}{2a} \div a^2$ (h) $\dfrac{a^2}{5b} \div a$

3 Simplify each of these.

(a) $\dfrac{5}{(x+1)^2} \times (x+1)$ (b) $\dfrac{1}{x} \div y$ (c) $\dfrac{45}{2\pi x} \times \dfrac{x^2}{9}$ (d) $\dfrac{x}{6} \div \dfrac{5}{x}$

(e) $\dfrac{x^3}{3y^2} \div \dfrac{x^2}{6y}$ (f) $(x-1) \times \dfrac{1}{x^2-1}$ (g) $\dfrac{2x+2}{x+4} \times \dfrac{x-3}{x+1}$ (h) $\dfrac{x^2-25}{12} \div \dfrac{x+5}{4}$

D Solving equations that contain fractions

1 Solve each of these equations.

(a) $\dfrac{3x+2}{x} = 4$ (b) $\dfrac{3x-1}{4} = x-1$ (c) $\dfrac{15-x}{4} = x$

(d) $\dfrac{x+8}{x+2} = 3$ (e) $\dfrac{x+1}{x} = 9$ (f) $\dfrac{2x-5}{x-2} = 4$

(g) $\dfrac{8x-3}{2x-1} = 6$ (h) $\dfrac{3(x+5)}{x+3} = 6$ (i) $\dfrac{2-3x}{2} = x+6$

2 Solve each of these equations.

(a) $\dfrac{6}{x-1} = x$ (b) $\dfrac{18}{x} = x+7$ (c) $\dfrac{7}{x-6} = x$

(d) $\dfrac{2x+5}{x} = x-2$ (e) $\dfrac{12}{x-3} = x+1$ (f) $\dfrac{x+10}{x} = 3x$

(g) $\dfrac{4(1-2x)}{x-6} = 2x-1$ (h) $\dfrac{3}{2x+1} = 2x-1$ (i) $\dfrac{2(6x-7)}{x-1} = 3x+2$

*3 Solve the equation $\dfrac{8x-1}{5x+9} = \dfrac{2x-1}{3x-1}$.

13 Time series

A Trend and moving average

1 The table below shows the quarterly ice cream sales for a small shop.

Year	2006			2007				2008
Quarter	2	3	4	1	2	3	4	1
Sales	450	850	160	93	380	880	145	86

(a) Calculate a 4-point moving average and show it with the original data on a graph.

(b) Describe the trend.

2 This table shows the amount spent on gas for a family of four.

Year	2006			2007				2008
Quarter	2	3	4	1	2	3	4	1
Amount (£)	65	38	195	112	55	42	172	109

(a) Calculate a 4-point moving average and show it with the original data on a graph.

(b) Describe the trend.

3 A council drop-in centre is open on weekdays only.
The table below shows the number of visitors during a period of two weeks.

Day	Mon	Tues	Wed	Thurs	Fri	Mon	Tues	Wed	Thurs	Fri
Number of visitors	19	34	43	41	50	16	36	44	49	56

(a) Calculate a 5-point moving average and show it with the original data on a graph.

(b) Describe the trend.

4 A college enrolls students each year for an art course.
The number of students who have enrolled over a period of ten years
is shown in the table below.

Year	1998	1999	2000	2001	2002	2003	2004	2005	2006	2007
Number of students	23	18	26	22	20	23	27	24	25	28

(a) Calculate a 3-point moving average.   (b) Describe the trend.

5 The table shows the number of road accidents recorded each year in a town.

Year	1994	1995	1996	1997	1998	1999	2000	2001	2002
Accidents	134	195	220	188	270	235	193	215	170

(a) Calculate a 5-point moving average.   (b) Describe the trend.

Mixed practice 2

1 What is the value of each expression when $k = {}^-5$?

(a) $2k + 3$ (b) $(k - 4)^2$ (c) $\dfrac{1 - 4k}{-7}$ (d) $\dfrac{2k}{5} - 3$

2 A row of 11 one-pound coins is 24.6 cm long.
How long is a row of 7 one-pound coins (to the nearest 0.1 cm)?

3 Find the nth term of the arithmetic sequence 3, 8, 13, 18, 23, …

4 Mrs Jones invests £250 in a savings account for her grandson.
At a compound interest rate of 4%, how much will there be in the account after 10 years?

5 Vinod has a picture that measures 4 cm by 6 cm.
He enlarges it so that its area is nine times as big.
What are the new dimensions of the picture?

6 Expand and simplify these.

(a) $(2x - 3)(x + 5)$ (b) $(4y - 1)^2$ (c) $(2c + d)^2$ (d) $(3k + 2m)(2k - 5m)$

7 This is a regular nine-sided polygon with two diagonals drawn.
Calculate the angles marked with letters.

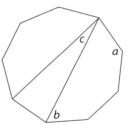

8 Solve these pairs of simultaneous equations.

(a) $4x + y = 11$ (b) $5x + 8y = 7$ (c) $2x - 3y = 17$
     $3x + 2y = 7$ $11x - 4y = 1$ $5x + 2y = 14$

9 One light-year is about 9.4×10^{12} km.
Alpha Centauri is the closest star to Earth. It is 4.34 light-years from Earth.
How far is it from Alpha Centauri to Earth in km, correct to two significant figures?

10 $4000 = 2^5 \times 5^n$.

(a) Find the value of n.

(b) Write 18×4000 as a product of prime factors.

11 Simplify each of these.

(a) $\dfrac{8n^2}{16n}$ (b) $\dfrac{5a + 15}{2a + 6}$ (c) $\dfrac{r^2 - 16}{3r - 12}$ (d) $\dfrac{10x^2 - 19x - 15}{25x^2 - 9}$

12 Find the radius of a circle with area 100 cm², correct to the nearest 0.1 cm.

13 This triangular prism has a volume of 1200 cm³.

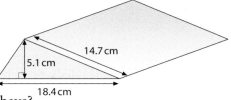

   (a) What is the length of the prism in cm, to 1 d.p?

   (b) Sketch a plan view of the prism.
   Mark lengths clearly on your sketch.

   (c) How many planes of symmetry does this prism have?

14 In one bag of red and blue counters the ratio of red to blue counters is $4:1$.
In another bag the ratio of red to blue counters is $4:5$.
A counter is chosen at random from each bag.
Find, as a fraction, the probability that the counters will be different colours.

15 The diagram shows a cuboid.

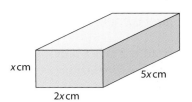

   (a) Find an expression for the volume of the cuboid.

   (b) The volume of the cuboid is 100 cm³.
   Find the area of the largest face, correct to 1 d.p.

16 Factorise these expressions fully.

   (a) $4x^2 - 8x - 5$ **(b)** $12a^2b + 8ab$ **(c)** $9x^2 - 6x + 1$ **(d)** $2x^2 - 50y^2$

17 There are 240 children in a school.
$\frac{1}{20}$ of the children are left-handed girls.
$\frac{1}{8}$ of the girls are left-handed.
$\frac{3}{8}$ of the left-handed children are girls.

	Girls	Boys
Left-handed		
Right-handed		

Copy and complete the two-way table to show the number in each section.

18 These candles are in the shape of square-based pyramids.
The larger candle is an enlargement of the smaller one.
The sloping edges have lengths 8 cm and 10 cm.
The volume of the smaller candle is 96 cm³.
What is the volume of the larger candle?

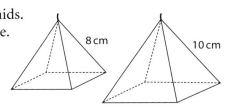

19 Q is directly proportional to $\sqrt{P}$.
When $P = 25$, $Q = 30$.

   (a) Find the equation connecting Q and P.

   (b) Find   **(i)** the value of Q when $P = 9$ **(ii)** the value of P when $Q = 60$

20 Greg wants to find the height of the tree in his garden.
He measures the angle of elevation of the top of the tree
as 64° at a distance of 5 m from the base of the tree.

Calculate the height of the tree, giving your answer
to an appropriate degree of accuracy.

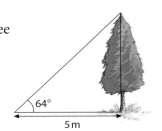

21 Solve these equations.

    (a) $9x^2 - 16 = 0$ **(b)** $2x^2 + 11x - 6 = 0$ **(c)** $3x^2 - 13x = 10$

22 A semicircle is removed from a quarter circle, as shown.
Prove that the shaded area remaining is
half the area of the original quarter circle.

23 In the following expressions the letters a, b, c and h all represent lengths.
π, 2 and 3 represent numbers that have no dimensions.
For each expression, state whether it could represent
a length, an area, a volume or none of these.

    (a) $3ab + ac$ **(b)** $2(a + b)$ **(c)** $\pi(a^2 + b^2)h$ **(d)** $a\sqrt{b^2 - c^2}$ **(e)** $\dfrac{a^2 + b^2}{\pi h}$

24 Brad is 5 years older than Angelina.
Cora is three times as old as Brad.
Their three ages add up to 80.
How old is Angelina?

25 The table below shows the takings of a garden centre for three-monthly periods.

Year	2006				2007			
Months	Jan–Mar	Apr–Jun	Jul–Sep	Oct–Dec	Jan–Mar	Apr–Jun	Jul–Sep	Oct–Dec
Takings (£000)	160	250	420	210	140	240	380	220

    (a) Calculate a four-point moving average for this data.

    (b) Comment on the trend in the takings.

26 Here are the first three three patterns in
a sequence of matchstick patterns.

    Pattern 1   Pattern 2   Pattern 3

    (a) How many matches are in the next pattern?

    (b) Find a rule for the number of matches in the nth pattern.

    (c) One of these matchstick patterns can be made using 419 matches.
         Which pattern is it?

27 The depth of the water in this cone is $\frac{2}{3}$ of the depth of the cone.

    What percentage of the volume of the cone does the water occupy?
    Give your answer correct to the nearest 1%.

28 Given that V is inversely proportional to the square of U, and $V = 10$ when $U = 2$,
find the value of V when $U = 4$.

29 Simplify each of these.

(a) $\dfrac{12a}{b^2} \times \dfrac{b}{8a}$
(b) $\dfrac{1}{b} \div a$
(c) $\dfrac{6ab^2}{a^4} \times \dfrac{a^2b}{2}$
(d) $\dfrac{a^2}{10} \div \dfrac{a}{5}$

30 Jackie makes a scale drawing of her garden.
She uses a scale of $1:50$.

(a) The real patio has an area of $22\,\text{m}^2$.
What area does it cover on the scale drawing?

(b) Jackie draws a flower bed with an area of $40\,\text{cm}^2$ on her scale drawing.
What is the area of the real flower bed?

(c) On the scale drawing her circular pond has a radius of $5\,\text{cm}$.
What is the area of the real pond?

31 The square and the triangle have the same area.
Find the value of a.

32 The mass of a solid rubber ball is proportional to the cube of the diameter.
A rubber ball of diameter $7.8\,\text{cm}$ has a mass of $0.96\,\text{kg}$.

Calculate, to an appropriate degree of accuracy,

(a) the mass in kg of a rubber ball of diameter $9.4\,\text{cm}$

(b) the diameter of a rubber ball that has a mass of $0.56\,\text{kg}$

33 This triangle is right-angled at B.

(a) Show that x must satisfy the equation $2x^2 - 15x + 7 = 0$.

(b) Solve the equation to find x.

(c) What is the exact value of sin C?

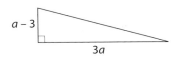

34 The expression $\dfrac{\pi p^2 q^n}{\sqrt{p^2 + q^2}}$ has the dimension of an area.

Find the value of n.

35 Factorise $2m^2 + 6mn + 4n^2$.

36 The first five terms of a sequence are $5,\ 12,\ 21,\ 32,\ 45,\ \ldots$
Find an expression for the nth term of the sequence.

37 North Houton has a population of 2845 with a mean age of 35.4.
South Houton has a population of 1225 with a mean age of 42.6.
Calculate the mean age of the combined population of North and South Houton.

38 Solve each of these equations.

(a) $\dfrac{4(x-3)}{2x} = 3$
(b) $\dfrac{3x+5}{2x-3} = 2$
(c) $\dfrac{10}{x+2} = 2x + 3$
(d) $\dfrac{2(x-1)}{x+3} = x - 4$

14 Trigonometric graphs

A The sine graph

1 Sketch the graph of $\sin x$ between $^-360°$ and $360°$.
Using a calculator, find these to the nearest $0.1°$,
giving all possible answers between $^-360°$ and $360°$.

   (a) $\sin^{-1} 0.3$ **(b)** $\sin^{-1} {}^-0.65$ **(c)** $\sin^{-1} 0.8$ **(d)** $\sin^{-1} 0.98$ **(e)** $\sin^{-1} {}^-0.34$

2 What angles between $^-360°$ and $360°$ have the same sine as each of these?

   (a) $45°$ **(b)** $^-120°$ **(c)** $280°$ **(d)** $^-205°$ **(e)** $390°$

3 Find all possible values of x between $0°$ and $360°$ for each of these.

   (a) $3 \sin x = 0.6$ **(b)** $5 \sin x = 3$ **(c)** $\frac{1}{2} \sin x = 0.35$ **(d)** $4 \sin x + 1 = 0$

B The cosine graph

1 Sketch the graph of $\cos x$ between $^-360°$ and $360°$.
Using a calculator, find these to the nearest $0.1°$,
giving all possible answers between $^-360°$ and $360°$.

   (a) $\cos^{-1} 0.8$ **(b)** $\cos^{-1} {}^-0.74$ **(c)** $\cos^{-1} {}^-0.3$ **(d)** $\cos^{-1} 0.05$ **(e)** $\cos^{-1} {}^-0.14$

2 What angles between $^-360°$ and $360°$ have the same cosine as each of these?

   (a) $10°$ **(b)** $100°$ **(c)** $^-50°$ **(d)** $^-342°$ **(e)** $450°$

3 Find all possible values of x between $0°$ and $360°$ for each of these.

   (a) $\frac{1}{2} \cos x = 0.49$ **(b)** $2 \cos x = {}^-1.4$ **(c)** $1 - \cos x = 0.65$ **(d)** $3 \cos x + \frac{1}{4} = 0$

*4 Solve $\cos x = \cos (5x)$, given that x is between $0°$ and $90°$.

C The tangent graph

1 Sketch the graph of x between $^-90°$ and $450°$.
Using a calculator, find these to the nearest $0.1°$,
giving all possible answers between $^-90°$ and $450°$.

   (a) $\tan^{-1} 0.7$ **(b)** $\tan^{-1} {}^-4$ **(c)** $\tan^{-1} 30$ **(d)** $\tan^{-1} {}^-0.2$ **(e)** $\tan^{-1} 0.05$

2 What angles between $^-90°$ and $450°$ have the same tangent as each of these?

   (a) $0°$ **(b)** $100°$ **(c)** $400°$ **(d)** $^-60°$ **(e)** $267°$

3 Find all possible values of x between $0°$ and $360°$ for each of these.

   (a) $\tan x = 3$ **(b)** $4 \tan x = {}^-1$ **(c)** $\frac{1}{2} \tan x = 0.65$ **(d)** $3 \tan x - 12 = 0$

D Graphs based on trigonometric functions

1 Sketch these graphs for x between $0°$ and $360°$, marking the key points.

   (a) $y = \frac{1}{2} \cos x$ **(b)** $y = \sin x + 3$ **(c)** $y = {}^-\sin x$ **(d)** $y = 2 \cos x - 1$

2 Suggest an equation for each of these graphs.

(a)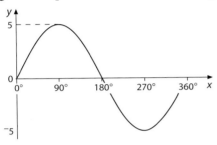

(b)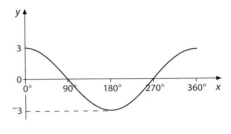

(c)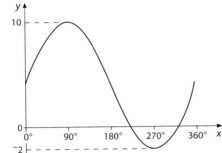

(d)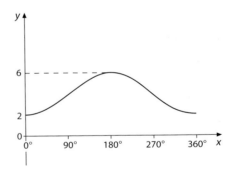

3 Sketch these graphs for x between $0°$ and $360°$, marking the key points.

   (a) $y = \sin(3x)$ **(b)** $y = 3 \cos(2x)$ **(c)** $y = \sin(4x) + 2$ **(d)** $y = \cos\left(\frac{1}{2}x\right)$

4 Suggest an equation for each of these graphs.

(a)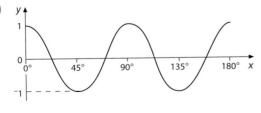

(b)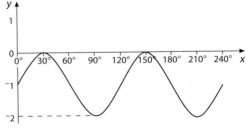

15 Pyramid, cone and sphere

This information is on the formula page that you will have in the GCSE Higher tier exam.

Volume of sphere = $\frac{4}{3}\pi r^3$

Surface area of sphere = $4\pi r^2$

Volume of cone = $\frac{1}{3}\pi r^2 h$

Curved surface area of cone = $\pi r l$

A Pyramid and cone

1 Calculate the volume of each of these pyramids.

(a)

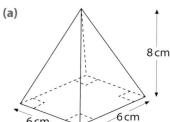

8 cm

6 cm   6 cm

(b)

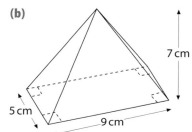

7 cm

5 cm   9 cm

(c)

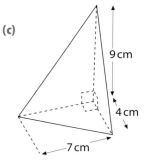

9 cm

4 cm

7 cm

2 A pyramid has a base that is a regular hexagon with sides 6.0 cm long.

  (a) Use trigonometry to find length p.

  (b) Find the total area of the hexagonal base.

  (c) The pyramid's height is 10.0 cm. Find its volume.

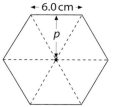

6.0 cm

p

3 This pyramid has a regular pentagon for its base.
Find the total area of the pyramid's five triangular faces.

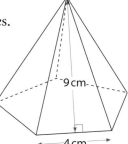

9 cm

4 cm

4 Find the volume and curved surface area of each of these cones to 3 s.f.

  (a) Height $h = 12$ cm, radius $r = 9$ cm, slant height $l = 15$ cm

  (b) Height = 21 cm, radius = 20 cm

  (c) Height = 35 cm, slant height = 37 cm

  (d) Diameter = 20 cm, height = 20 cm

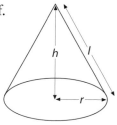

h l

r

5 This is a frustum of a cone. Find, to 3 s.f.,

(a) its volume

(b) the area of its curved surface

(c) its total surface area, including the circular top and base

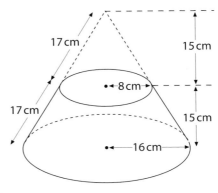

*6 A cone of volume 5000 cm³ has a radius of 8.0 cm. Find the curved surface area of the cone to 3 s.f.

B Sphere

1 Calculate the volume and surface area of spheres with these radii, giving your answers to 3 s.f.

(a) 4.0 cm   **(b)** 15.0 cm   **(c)** 2.5 m   **(d)** 11.0 mm

2 A tank for fuel is a sphere with radius 1.2 m. How many litres of fuel can it hold?

3 Calculate the volume of wood in this rolling pin with hemispherical ends.

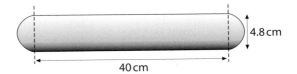

4 Find the total mass of 100 steel ball bearings, each with a diameter of 12 mm, given that the density of steel is 7.8 g/cm³.

5 The Earth is approximately a sphere with radius 6.37×10^3 km. Find its surface area and volume, giving your answers in standard form.

6 The glass tank with the measurements shown is half filled with water. A metal sphere with radius 4.0 cm is placed in the tank. How far will the water level rise?

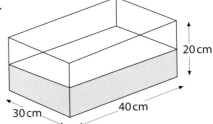

16 Rearranging a formula

A Review: rearranging a simple formula

B Choosing the first step in a rearrangement

1 The equation of a straight line is $x + 3y = 2$.
- (a) (i) Rearrange it to make x the subject.
  - (ii) Where does the line cross the x-axis?
- (b) (i) Rearrange the equation to make y the subject.
  - (ii) What is the gradient of the line?

2 Rearrange each of these formulas so that the bold letter is the subject.
- (a) $p = 5\boldsymbol{q} - 8$
- (b) $2\boldsymbol{x} - 3y = 8$
- (c) $h = 10 - 3\boldsymbol{k}$
- (d) $m = \dfrac{\boldsymbol{n} - 4}{7}$
- (e) $p = \dfrac{\boldsymbol{q}}{3} + 2$
- (f) $a = \dfrac{4\boldsymbol{b}}{9} - 7$
- (g) $y = \dfrac{8 - \boldsymbol{x}}{3}$
- (h) $g = \dfrac{\boldsymbol{h}}{k} - 1$
- (i) $y = 5 - x\boldsymbol{z}$

3 Rearrange each of these formulas so that the bold letter is the subject.
- (a) $p = 5(\boldsymbol{q} - 7)$
- (b) $f = 5(7 - \boldsymbol{g})$
- (c) $g = h(\boldsymbol{k} - 4)$
- (d) $g = \boldsymbol{h}(k - 4)$
- (e) $a = 5\boldsymbol{b}c$
- (f) $a = 5b\boldsymbol{c}$
- (g) $2w = \boldsymbol{x}(6 - y)$
- (h) $2w = x(6 - \boldsymbol{y})$
- (i) $d = 3\boldsymbol{ef} + g$
- (j) $r = \pi t^2\boldsymbol{s}$
- (k) $R = 4r^2\boldsymbol{d} - t^2$
- (l) $p = q^2 - \boldsymbol{st}$

4 Make z the subject of each of these formulas.
- (a) $v = \frac{1}{5}z + 3$
- (b) $y = \dfrac{z + 12}{3}$
- (c) $x = \frac{1}{4}(z - 20)$
- (d) $y = \dfrac{x}{z} + 1$
- (e) $m = \dfrac{kzn}{3} - 2$
- (f) $f = \frac{1}{8}g^2z$
- (g) $y = \dfrac{2x}{z}$
- (h) $y = \dfrac{\pi xz + 9}{5}$
- (i) $y = \dfrac{\pi}{xz}$
- (j) $r = q - \dfrac{p}{z}$
- (k) $y = \dfrac{4}{x + z}$
- (l) $y = \dfrac{w + 1}{z - 2}$
- (m) $2x = \dfrac{5}{y - z}$
- (n) $m = pn - \dfrac{3\pi}{yz}$
- (o) $w - 7 = \dfrac{\pi u}{v - z}$

5 The volume of a rectangular pyramid is given by $V = \dfrac{abh}{3}$.
- (a) Rearrange the formula to make h the subject.
- (b) Calculate the height of a rectangular pyramid
  of volume $96\,\text{cm}^3$, where $a = 4\,\text{cm}$ and $b = 6\,\text{cm}$.

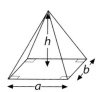

C Simplifying squares and square roots
D Formulas involving squares and square roots

1 Multiply out the brackets and simplify each expression.

(a) $(3x)^2$ (b) $(4x^2y)^2$ (c) $\left(\frac{1}{4}xy^3\right)^2$ (d) $\left(\frac{x}{5b}\right)^2$ (e) $\left(\frac{3y}{7x}\right)^2$

(f) $\left(\sqrt{x}\right)^2$ (g) $\left(5\sqrt{p}\right)^2$ (h) $\left(2x\sqrt{y}\right)^2$ (i) $\left(\frac{\sqrt{8}}{x}\right)^2$ (j) $\left(\frac{\sqrt{y}}{x\sqrt{7}}\right)^2$

2 Simplify each expression.

(a) $\sqrt{9x^2}$ (b) $\sqrt{144x^6}$ (c) $\sqrt{x^4y^2}$ (d) $\sqrt{\frac{1}{9}x^2}$ (e) $\sqrt{\frac{x}{16}}$

3 In each of these formulas r is a length.
   Rearrange each of these to make r the subject.

(a) $A = 25r^2 - 1$ (b) $t = \frac{7r^2}{12}$ (c) $S = \frac{1}{4}\pi r^2$ (d) $3r^2 + h^2 = 30$

(e) $V = \frac{2\pi r^2 l}{3}$ (f) $A = 2r^2\pi + 2p^2$ (g) $H = 7\pi r^2 + 6\pi y^2$ (h) $2h = \frac{r^2}{3\pi ab}$

4 The power, in watts, dissipated in an electric circuit is given by $W = I^2R$.
   I stands for the current in amps and R is the resistance of the circuit in ohms.

(a) Rearrange the formula to give I in terms of W and R.

(b) Calculate the current in an electric circuit where $W = 150$ watts and $R = 12.25$ ohms.

5 Make x the subject of each of these formulas.

(a) $y = \sqrt{x - 3}$ (b) $y = \sqrt{x} - 3$ (c) $y = \sqrt{5x}$ (d) $y = 5\sqrt{x}$

(e) $y = \sqrt{\frac{x}{z}}$ (f) $y = \frac{\sqrt{x}}{z}$ (g) $y = 2\pi\sqrt{\frac{x}{v}}$ (h) $y = \sqrt{\frac{2\pi v}{x}}$

(i) $y = \sqrt{k - 5x}$ (j) $y = \frac{\sqrt{x}}{3k}$ (k) $y = \frac{1}{9}\pi k\sqrt{x - z}$ (l) $y = \pi r\sqrt{x + k^2}$

6 In each of these formulas, a, b and c can take positive and negative values.
   Rearrange each one to make b the subject.

(a) $a = 2b^2 + 15$ (b) $a^2 + b^2 = c^2$ (c) $c = \frac{a - b^2}{6\pi}$ (d) $a = \frac{c^2}{b^2}$

E Formulas where the new subject appears more than once

1 Make k the subject of each of these formulas.

(a) $g = 3k + xk$ (b) $h = 2k - jk$ (c) $f = gk + hk$

(d) $pk = 4(k + 1)$ (e) $2k = a(k + 3)$ (f) $k + h = \pi kV^2$

2 Rearrange each of these to make x the subject.

    (a) $4(x - 3l) = 2x + 5$ **(b)** $xz - a = xy + b$ **(c)** $2x = z(x + s)$

    (d) $\pi x^2 + 2x^2 = yz$ **(e)** $P = 2x + \pi x - 9$ **(f)** $\pi(x - 9) = z(x + 2)$

    (g) $H = 3(x - 2) + \pi x$ **(h)** $a(x - 3) = y(10 - 3x)$ **(i)** $x^2 y^2 = \frac{1}{2}x^2 + 3$

3 Make k the subject of each of these.

    (a) $g = \dfrac{k + 12h}{k}$ **(b)** $g = \dfrac{3h - k}{k}$ **(c)** $g = \dfrac{4h + 5k}{8k}$

4 Rearrange each of these to make y the subject.

    (a) $P = \dfrac{2y}{y - 1}$ **(b)** $H = \dfrac{4 - y}{y - 3}$ **(c)** $x = \dfrac{2(y + 3)}{y + z}$

    (d) $I = \dfrac{3(x - 2y)}{y + 2x}$ **(e)** $q = \dfrac{p^2 + y}{p - y}$ **(f)** $r = 3y + \dfrac{\pi xy}{40}$

5 Rearrange $a = \dfrac{4(b + c)}{bc}$ to make b the subject.

F Forming and manipulating formulas

1 The diagram shows a trapezium.

    (a) Find the formula for the area A of the trapezium.

    (b) Rearrange the formula to make x the subject.

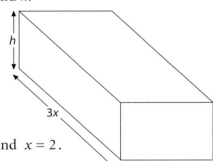

    (c) What are the dimensions of this trapezium when its area is $75\,\text{cm}^2$?

    (d) Find the height of this trapezium (to the nearest 0.1 cm) when its area is $400\,\text{cm}^2$.

2 The box shown is a cuboid with edges of length x, $3x$ and h.

    (a) Find the formula for its volume V.

    (b) Rearrange this formula to make x the subject.

    (c) Work out the value of x for a box like this with a volume of $1000\,\text{cm}^3$ and a height of 5 cm.

    (d) Find the formula for the surface area A of the box.

    (e) Rearrange this formula to make h the subject.

    (f) Work out the value of h for a box where $A = 88$ and $x = 2$.

3 The diagram shows a sector of a circle with radius r.
It is folded up to make a cone.

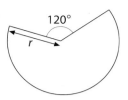

(a) Find the formula for S, the curved surface area of the cone.

(b) Rearrange this formula to make r the subject.

4 The diagram shows a waste bin in the shape of
a hemisphere of radius r cm on top of a cylinder of height $2r$ cm.

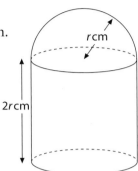

(a) Show that the total volume V cm^3 of the bin
is given by $V = \frac{8}{3}\pi r^3$.

(b) Rearrange this formula to make r the subject.

(c) The company wishes to make waste bins with
a total volume of 35 litres.
Calculate the value of r for this size bin.

Give your answer to a sensible degree of accuracy.

5 A pyramid has a square base with edge length r and height h.
A sphere has radius $\frac{1}{2}r$.

(a) What is the volume of the pyramid
in terms of r and h?

(b) Find an expression for the volume
of the sphere in terms of r.

The pyramid and the sphere have
the same volume.

(c) Show that $h = \frac{1}{2}\pi r$.

6 The diagram shows a table top made of
glass with a metal edge.
The shape of the glass top is formed from
a square and two semicircles, each with
a radius of r cm.
The metal edge is a strip with a constant
width of w cm.

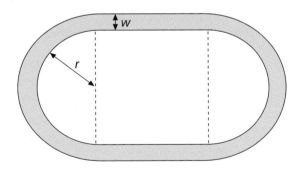

The area of the top surface of the metal
strip is A cm^2.

(a) Find a formula for A in terms of r and w.

(b) Show that the formula can be rearranged to give $r = \dfrac{A - \pi x^2}{2w(\pi + 2)}$.

(c) A table is to be made so that $A = 2000$ and $w = 8$.

(i) Find the corresponding value of r.

(ii) What will be the width of this table?

17 Sampling

A Representative samples
B Random sampling

1 Janice works in a library. She wants to estimate how often, on average, a book is taken out.

She looks at 50 books that have been returned today to see how many times each book has been issued in the past year.

Explain why Janice's sample of 50 books is not representative of all the books in the library.

2 Patrick finds an old family photo album with 120 pages. The number of photos on a page varies. Patrick wants to estimate the number of photos in the album.

He numbers the pages from 1 to 120 and chooses a random sample of 20 pages. He counts the photos on each page of his sample, with these results:

6 8 5 6 6 7 8 4 7 5 8 3 7 6 6 7 5 2 6 4

Estimate the total number of photos in the album.

D Stratified sampling

Stratified sampling may be assumed to be proportional.

1 A college has 450 full-time students and 190 part-time students.
A stratified sample of 100 students is to be selected.
How many of each type of student should be included?

2 This table shows the numbers of students in a school broken down into lower school (years 7–9), upper school (years 10–11) and sixth form.

A stratified sample of 50 students is to be selected. How many students from each part of the school should be included?

	Number of students
Lower school	363
Upper school	273
Sixth form	195

3 This table gives information about the members of a club.

A stratified sample of 80 members is to be selected. How many of the members in the sample should be

	60 or under	Over 60
Male	72	29
Female	53	38

(a) male   (b) over 60   (c) females over 60

18 Using graphs to solve equations

You need graph paper.

> A **Review: graphs and simple equations**
> B **Intersection of a linear and a curved graph**

1 (a) Copy and complete the table below for $y = x^2 - 2x - 4$.

x	$^-3$	$^-2$	$^-1$	0	1	2	3	4	5
y	11			$^-4$					

(b) On a grid like this, draw the graph of $y = x^2 - 2x - 4$.

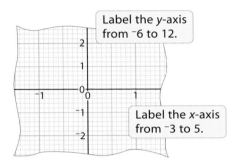

Label the y-axis from $^-6$ to 12.

Label the x-axis from $^-3$ to 5.

(c) Use your graph to estimate the values of x for which

(i) $x^2 - 2x - 4 = 0$ **(ii)** $x^2 - 2x - 4 = 10$

(d) Explain how your graph shows that the equation $x^2 - 2x - 4 = ^-6$ has no solution.

(e) (i) By drawing a suitable line on the graph, estimate the solutions to the equation $x^2 - 2x - 4 = 2x - 3$.

(ii) Show that $x^2 - 2x - 4 = 2x - 3$ can be simplified to $x^2 - 4x - 1 = 0$.

2 The curve shown is that of $y = x^3 - 3x^2 + 3$. The straight line is $y = x - 1$.

(a) Use the graphs to estimate the solutions to the equation $x^3 - 3x^2 + 3 = x - 1$.

(b) (i) Show that these solutions will satisfy the equation $x^3 - 3x^2 - x + 4 = 0$.

(ii) Use trial and improvement to find the largest solution to this equation, correct to two decimal places.

(c) Explain how the graph shows that $x^3 - 3x^2 + 3 = ^-2$ has only one solution.

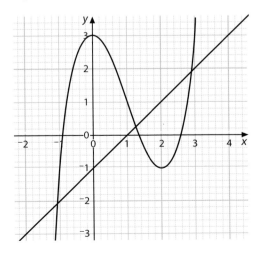

C Deciding which linear graph to draw

1 Which of these are rearrangements of the equation $x^2 - 7x + 3 = 0$?

A $x^2 - 8x + 5 = 2 - x$

B $x^2 - 6x + 1 = 2 - x$

C $2x^2 + 3 = x^2 + 7x$

D $x^2 - 6x + 1 = x - 2$

E $x^2 - 6x + 5 = x + 2$

2 If you have the graph of $y = x^2 + 5x$, which of these straight lines do you need to draw to estimate the solutions to the equation $x^2 - x + 1 = 0$?

A $y = 6x + 1$

B $y = 6x - 1$

C $y = 1 - 6x$

D $y = 4x - 1$

E $y = 4x + 1$

3 If you have the graph of $y = x^2 + 3x - 7$, what straight line do you need to draw to use it to estimate the solutions to the equation $x^2 + x - 7 = 0$?

4 If you have the graph of $y = 2x^2 + x$, what straight line do you need to draw to use it to estimate the solutions to the equation $2x^2 + 6x - 1 = 0$?

5 (a) Copy and complete the table below for $y = x^3 - 2x$.

x	-2	-1.5	-1	-0.5	0	0.5	1	1.5	2
y	-4					-0.875			

(b) On graph paper, draw the graph of $y = x^3 - 2x$ for $-2 \leq x \leq 2$.

(c) (i) Use the graph to show that the equation $x^3 - 2x - 3 = 0$ has only one solution.

(ii) Estimate the solution to $x^3 - 2x - 3 = 0$.

(iii) Use trial and improvement to find this solution correct to two decimal places.

(d) By drawing a suitable line on the graph, estimate all the solutions to the equation $x^3 - 3x - 1 = 0$.

6 (a) On graph paper, draw and label axes with $-3 \leq x \leq 3$ and $-12 \leq y \leq 12$.
On your axes, draw the graph of $y = \dfrac{12}{x}$.

(b) By drawing a suitable line on the graph, estimate all the solutions to the equation $\dfrac{12}{x} - 3x - 1 = 0$.

(c) Draw the line $y = 3 - x$ on your graph.
Use the line to show that there is no solution to the equation $\dfrac{12}{x} + x - 3 = 0$.

7 If you have the graph of $y = x^2 - 2x$, what **straight** line do you need to draw to estimate the solutions to the equation $3x^2 - 7x - 3 = 0$?

8 If you have the graph of $y = 4x^2 + 1$, what straight line do you need to draw to estimate the solutions to the equation $x^2 - 5x + 1 = 0$?

1 A rectangular camp site is to be built alongside a river.
   The width of the site is w metres.

   The site is to have an area of $4000\,\text{m}^2$.

   (a) Show that the length of fencing needed, $L\,\text{m}$,
       is given by the formula $L = 2w + \dfrac{4000}{w}$.

   (b) Draw the graph of L against w for $10 \leq w \leq 200$.

   (c) A site like this is made with $300\,\text{m}$ of fencing.
       Use your graph to estimate the two possible
       values for its width.

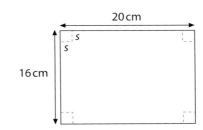

   (d) Estimate the **minimum** length of fencing needed to make a site like this.
       What is the approximate width of a site that uses this minimum length of fencing?

2 A small, open gift box is to be made from
   a piece of card measuring $16\,\text{cm}$ by $20\,\text{cm}$.
   A square of side $s\,\text{cm}$ is cut from each corner.

   The card is then folded up to make the box.
   Its capacity is $C\,\text{cm}^3$.

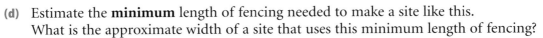

   (a) Find a formula for C in terms of s.

   (b) Draw the graph of C against s for
       a sensible range of values for s.

   (c) (i) Estimate the value of s that gives the box with the largest possible capacity.

       (ii) The volume of a large tube of chocolate beans is $400\,\text{cm}^3$.
            Explain, with reasons, whether you think the beans would fit in this gift box.

3 A hollow cuboid is to be made from wood and all its surfaces painted gold.
   The square hole goes all the way through.

   The solid is to have a volume of $640\,\text{cm}^3$ and
   the area painted gold is to be $580\,\text{cm}^2$.

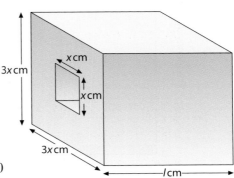

   (a) Show that $l = \dfrac{80}{x^2}$.

   (b) Find an expression, in terms of l and x,
       for the area to be painted gold.

   (c) Show that x satisfies the equation
       $4x^3 - 145x + 320 = 0$.

   (d) (i) Draw the graph of $y = 4x^3 - 145x + 320$
            for $0 \leq x \leq 6$.

       (ii) Hence estimate the dimensions (to the nearest $0.1\,\text{cm}$) of two different
            solids that fit the requirements for the volume and painted area.

19 Histograms

You need squared paper.

A Reading a histogram using an 'area scale'
B Frequency density

1 A company surveyed its employees to find out how far they each travelled to work.
   The results are summarised in the table below.

Distance, d km	Frequency
$0 < d \leq 5$	10
$5 < d \leq 10$	22
$10 < d \leq 20$	15
$20 < d \leq 50$	6
$50 < d$	0

[ ] represents ... employees

Distance (km)

The bar for $0 < d \leq 5$ has been drawn.

(a) What does one square on the histogram represent?

(b) Copy and complete the histogram.

2 A survey of a group of schoolchildren was carried out to find out
   how much pocket money they received each week.
   The distribution is shown in this histogram.

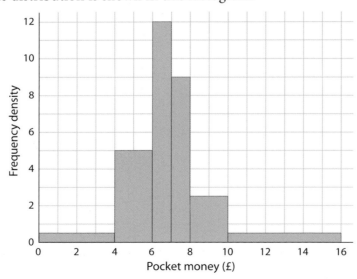

Frequency density

Pocket money (£)

(a) How many children received less than £4 each week?

(b) How many children received more than £8 each week?

(c) How many children were surveyed altogether?

3 Copy and complete the frequency table below, using the information in the histogram.

Weight, w g	Frequency
20 < w ≤ 40	
40 < w ≤ 50	
50 < w ≤ 60	
60 < w ≤ 80	

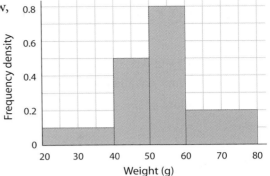

c Drawing a histogram

1 This table shows the distribution of the weights of a group of 15-year-old boys.

Weight, w kg	30 < w ≤ 40	40 < w ≤ 50	50 < w ≤ 55	55 < w ≤ 60	60 < w ≤ 70	70 < w ≤ 90
Frequency	2	10	15	13	6	4

(a) Calculate the frequency density for each interval.

(b) Draw a histogram for the data.

2 The unfinished histogram and table give information about the length of time taken by some members of a health club to run one mile.

Copy and complete the table and histogram.

Time, t minutes	Frequency
5 < t ≤ 8	
8 < t ≤ 9	
9 < t ≤ 10	13
10 < t ≤ 12	8
12 < t ≤ 14	2

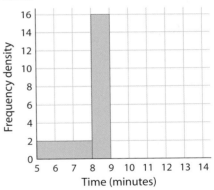

3 Draw a histogram for this data about the ages of cars in a car park.

Age, A years	0 < A ≤ 1	1 < A ≤ 3	3 < A ≤ 6	6 < A ≤ 10	10 < A ≤ 20
Frequency	6	27	15	14	3

4 Draw a histogram for this data about the salaries of the employees of a company.

Salary (thousand £)	10–15	15–25	25–40	40–60	60–90
Number of employees	60	110	75	30	15

20 Quadratic expressions and equations 2

This information is on the formula page that you will have in the GCSE Higher tier exam.

The quadratic equation

The solutions of $ax^2 + bx + c = 0$, where $a \neq 0$, are given by $x = \dfrac{-b \pm \sqrt{b^2 - 4ac}}{2a}$

A Perfect squares
B Using a prefect square to solve a quadratic equation

1 Copy and complete these identities.

   (a) $(x + \blacksquare)^2 = x^2 + 6x + \blacksquare$ **(b)** $(x - \blacksquare)^2 = x^2 - 8x + \blacksquare$

   (c) $(x + \blacksquare)^2 = x^2 + 12x + \blacksquare$ **(d)** $(x - \blacksquare)^2 = x^2 - 10x + \blacksquare$

2 Which of these expressions are perfect squares?

   A $x^2 + 4x + 4$ **B** $x^2 + 2x + 2$ **C** $x^2 + 14x + 49$

   D $x^2 - 6x + 9$ **E** $x^2 - 16x + 32$ **F** $x^2 - 12x - 36$

3 What number do you need to add to each expression to make it a perfect square?

   (a) $x^2 + 8x$ **(b)** $x^2 - 4x$ **(c)** $x^2 + 10x + 15$

   (d) $x^2 + 2x - 3$ **(e)** $x^2 - 16x + 20$ **(f)** $x^2 - 2x - 7$

4 What number do you need to subtract from each expression to make it a perfect square?

   (a) $x^2 + 4x + 5$ **(b)** $x^2 + 10x + 40$ **(c)** $x^2 - 6x + 15$

In questions 5–7, give solutions to three decimal places where appropriate.

5 Solve these equations.

   (a) $(x + 2)^2 = 36$ **(b)** $(x - 6)^2 = 49$ **(c)** $(x - 1)^2 = 2$

6 Solve these equations by using perfect squares.

   (a) $x^2 + 4x - 6 = 0$ **(b)** $x^2 - 8x - 12 = 0$ **(c)** $x^2 + 10x + 8 = 0$

   (d) $x^2 + 2x - 13 = 0$ **(e)** $x^2 + 14x + 3 = 0$ **(f)** $x^2 - 20x + 15 = 0$

7 **(a)** Expand and simplify $\left(x + \frac{9}{2}\right)^2$.

   (b) Hence solve the equation $x^2 + 9x - 4 = 0$.

   (c) Solve the equation $x^2 + 5x - 8 = 0$ by using a perfect square.

c Using a formula to solve a quadratic equation

In these questions, give solutions to three decimal places where appropriate.

1 Use the quadratic formula to solve each equation.

(a) $x^2 + 3x - 2 = 0$ (b) $x^2 - 6x + 2 = 0$ (c) $2x^2 - 5x + 3 = 0$

(d) $3x^2 + 4x - 8 = 0$ (e) $9x^2 = 12x + 4$ (f) $8x = x^2 + 5$

2 Solve these equations, choosing your own method each time.

(a) $x^2 - 7x + 12 = 0$ (b) $x^2 + 8x - 3 = 0$ (c) $6x^2 + 4x - 2 = 0$

(d) $5x^2 - 3x - 4 = 0$ (e) $x^2 + 2x = 24$ (f) $4(x^2 + 1) = 10x$

3 (a) Solve the equation $x^2 - 10x + 20 = 0$.

(b) Use your solution to choose the correct sketch of $y = x^2 - 10x + 20$.

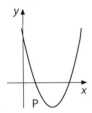 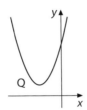 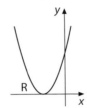 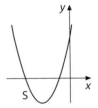

4 (a) What happens when you use the formula to solve $x^2 - 5x + 20$?

(b) What does this tell you about the graph of $y = x^2 - 5x + 20$?

D Solving problems

1

(2x + 3) cm

rectangle A $2x$ cm

(x − 1) cm

rectangle B   (x − 2) cm

The area of rectangle A is $55\,\text{cm}^2$ greater than the area of rectangle B.

(a) Show that $x^2 + 3x - 19 = 0$.

(b) By solving the equation $x^2 + 3x - 19 = 0$, find the value of x correct to 3 d.p.

2 A rectangular pond has a path on four sides as shown.
The pond has dimensions x metres by $4x$ metres.
The path is 2 metres wide.
The total area of the pond and the path is $91\,\text{m}^2$.

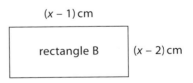

$4x$ m

pond x m

(a) Show that $4x^2 + 20x - 75 = 0$.

(b) By solving the equation $4x^2 + 20x - 75 = 0$, find the area of the pond.

3 The area of this triangle is $20\,\text{cm}^2$.

    (a) Show that $x^2 + 5x - 40 = 0$.

    (b) Solve the equation $x^2 + 5x - 40 = 0$.
        Hence find the length of QR, correct to 2 d.p.

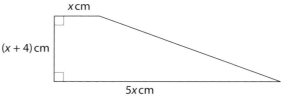

4 The area of this shape is $90\,\text{cm}^2$.

    (a) Show that $x^2 + 4x = 30$.

    (b) Find the value of x, correct
        to three significant figures.

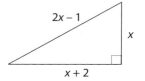

5 This is a right-angled triangle with dimensions as shown.
   Find the value of x, correct to three decimal places.

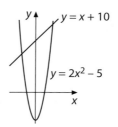

E Simultaneous equations

1 Solve these pairs of simultaneous equations, correct to 2 d.p. where appropriate.

   (a) $y = 10x - 16$
        $y = x^2$

   (b) $y = x^2 - 3$
        $y = x + 9$

   (c) $y = 2x^2 + x - 3$
        $y = 2x$

   (d) $y - 2x = 5$
        $y = 3x^2$

   (e) $y = 4x^2 - 3$
        $y + x = 9$

   (f) $y - 3x = 1$
        $y = 2(4 - x^2)$

2 Find the points of intersection for each pair of graphs.

   (a)

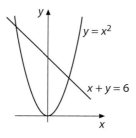

   (b)

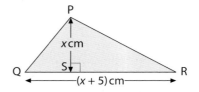

3 (a) Solve the equation $x^2 + 1 = 3x + 7$.

   (b) What does your solution show about the line $y = 3x + 7$ and the curve $y = x^2 + 1$?

4 (a) Show that the line $y = 7x - 1$ touches the curve $y = 9x^2 + x$ at one point only.

   (b) What are the coordinates of this point?

5 Show that the line $y = x - 1$ does not meet the curve $y = x^2 + 4x + 3$.

Mixed practice 3

You need graph paper.

1 Find the value of $3x - 2z$ when $x = {}^-5$ and $z = {}^-3$.

2 Find the value of each of these as an integer or fraction.

(a) $5^7 \div 5^5$
(b) $\dfrac{5^6 \times 5^5}{5^{12}}$
(c) 8^0
(d) 6^{-2}

3 (a) Work out an estimate for $\dfrac{20.5 \times 28.6}{6.1 \times 4.95}$.

(b) Use your answer to part (a) to find an estimate for $\dfrac{205 \times 2860}{610 \times 495}$.

4 Fionn measured the height in centimetres of 15 seedlings.

9.4 6.6 7.2 4.8 7.6 7.9 9.1 6.9 2.3 5.0 6.3 4.5 5.2 8.6 9.6

(a) Put this information into an ordered stem-and-leaf diagram. Include a key.

(b) What is the median length of these seedlings?

5 Rearrange each of these formulas to make the bold letter the subject.

(a) $V = \frac{1}{3}r\mathbf{s}t$
(b) $p = 2(q - \mathbf{r})$
(c) $a = \mathbf{b}(3 - c)$
(d) $y = \dfrac{2\mathbf{x}}{5} + z$
(e) $m = \dfrac{\pi}{p\mathbf{n}}$
(f) $p = \dfrac{2\pi}{r + \mathbf{q}}$

6 This table shows the age distribution of the members of a club.

Age group	15–30	30–45	45–60	60+
Number	76	62	31	11

Reena wants to take a sample of 40 of the members, stratified proportionally by age group. How many from each age group should be in the sample?

7 The diagram shows a prism.

(a) How many planes of symmetry does it have?

(b) Which one of these expressions gives its volume?

$\boxed{3(a^2 + b^2)}$ $\boxed{3(a^2 + b)}$ $\boxed{3a^2b^2}$ $\boxed{3a^2b}$

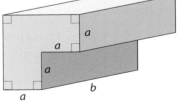

(c) (i) Show that the surface area S of the prism is given by the formula

$$S = 2a(3a + 4b)$$

(ii) Rearrange this formula to make b the subject.

8 Find all the possible values of x between $0°$ and $360°$ for each of these. Give values correct to the nearest $0.1°$ where necessary.

(a) $\sin x = 0.5$
(b) $\cos x = {}^-0.2$
(c) $\cos x + 1 = 0$
(d) $5 \sin x = {}^-3.6$

9 A child's toy consists of a cone, a cylinder and a hemisphere joined together as shown.

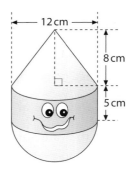

   (a) Calculate the total volume of the toy.
   Show all your working carefully.
   Give your answer to an appropriate degree of accuracy.

   (b) Do the same for the total surface area of the toy.

10 (a) On graph paper draw an accurate graph of $y = x^3 - 3x - 1$.
   Use values of x from $^-2$ to 3.

   (b) (i) What straight line can you draw to estimate the solutions to $x^3 - 4x - 2 = 0$?

   (ii) Draw this line and estimate the solutions.

11 This histogram shows the distribution of consultation times in a medical practice one day.

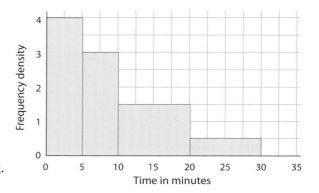

   (a) How many consultations lasted no more than 5 minutes?

   (b) What percentage of consultations lasted 10 minutes or more?

   (c) Estimate the mean length of a consultation, showing your working.

12 A strip of width x cm is cut from each edge of a rectangle whose sides were originally of length 10 cm and 7 cm.

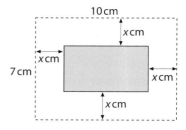

   (a) Write down an expression for the area of the remaining rectangle.

   (b) Given that the area remaining is 50 cm^2, calculate the value of x, correct to 1 d.p.

13 This vase is a frustum of a cone.
Calculate its volume.

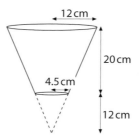

14 (a) The expression $x^2 - 14x + a$ can be written in the form $(x - b)^2$.
   Find the values of a and b.

   (b) Solve the equation $x^2 - 14x + 15 = 0$, giving your answers to 3 d.p.

15 Find all the possible values of x between $^-180°$ and $180°$ for which $\tan x = 1$.

16 In each of these formulas x can take positive and negative values.
Make x the subject of each formula.

   (a) $y = \frac{1}{5}\sqrt{x}$ **(b)** $y = \sqrt{x - 5\pi}$ **(c)** $y = z - \sqrt{wx}$

   (d) $V = 2\pi\sqrt{\dfrac{x}{5z}}$ **(e)** $y^2 = \dfrac{x^2}{5}$ **(f)** $A = \pi r\sqrt{h^2 + x^2}$

17 This is the graph of
$y = \sin x$ for $0° \leq x \leq 360°$.

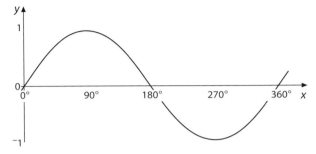

   (a) One solution to the equation $\sin x = ^-0.3$ is $x = 197°$ to the nearest degree.
Write down the other solution to the equation in the range $0° \leq x \leq 360°$.

   (b) On separate diagrams sketch the graphs of

      (i) $y = \sin x + 2$ **(ii)** $y = ^-\sin x$ **(iii)** $y = 2\sin 2x$

18 Solve these quadratic equations, giving your answers to 2 d.p. where necessary.

   (a) $3x^2 + 4x - 15 = 0$ **(b)** $2x^2 + 13x = 24$ **(c)** $4x^2 = 3x + 9$

19 The sides of a rectangle are of length a cm and b cm.
The perimeter is P cm and the area is A cm^2.
The length of the diagonal of the rectangle is d cm.

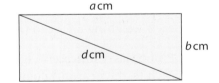

   (a) Show that $P^2 = 4d^2 + 8A$.

   (b) Rearrange this formula to make d the subject.

   (c) Find the length of the diagonal of a rectangle with area 10 cm^2 and perimeter 16 cm.

20 A cone has base radius 7 cm and curved surface area 200 cm^2.

   (a) Find the slant height of the cone.

   (b) Find the volume of the cone.

21 Make q the subject of each formula.

   (a) $p = \pi q + 2q$ **(b)** $p + q = pq$ **(c)** $R = \pi q - 3q + 5p$

   (d) $m = \dfrac{3 - q}{q}$ **(e)** $z = \dfrac{2q - 5}{q + 4}$ **(f)** $y = \dfrac{z^2 + q}{\pi + 3q}$

22 Solve the simultaneous equations $y - x = 5$ and $y = 2x^2 - 3$.
Give your solutions to 1 d.p.

23 What is the value of $\left(\frac{1}{3}\right)^{-3}$?

21 Working with rounded quantities

A Lower and upper bounds

1 What are the lower and upper bounds of the interval containing
all the numbers for which

(a) 6.4 is the nearest tenth   (b) 0.1 is the nearest tenth

(c) 50 is the nearest whole number   (d) 250 is the nearest ten

(e) 4900 is the nearest hundred   (f) 7.60 is the nearest hundredth

2 Write down the lower and upper bounds for each of these.

(a) The length of a room is 2.8 m, correct to two significant figures.

(b) The weight of a baby is 4.20 kg, correct to three significant figures.

(c) The volume of a pond is 3.1 m³, correct to two significant figures.

(d) The length of a work surface is 1370 mm, correct to the nearest 10 mm.

(e) The capacity of a jug is 350 ml, correct to the nearest 10 ml.

(f) The weight of a parcel to the nearest 100 g is 1.5 kg.

B Calculating with rounded quantities

1 Find the upper and lower bounds of the total weight of three boxes
each weighing 5.7 kg to the nearest 0.1 kg.

2 Philip has a suitcase weighing 18 kg and a rucksack weighing 4 kg,
each to the nearest kilogram.
Find the greatest and least possible weights of Philip's luggage.

3 The dimensions of a rectangular room are measured as 3.7 m and 2.1 m,
correct to two significant figures.
Find the upper and lower bounds of

(a) the perimeter of the room   (b) the area of the room

4 Lara records the distance she cycles each day for a week.
On five days she cycles 35 km and on two days she cycles 16 km.
The distances are measured to the nearest kilometre.
Find the upper and lower bounds of the distance she cycled during the week.

5 The dimensions of this trapezium are given to
the nearest centimetre.
Calculate the maximum possible area of the trapezium.

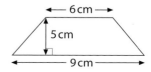

6 An aircraft travels at 460 m.p.h. for 9 hours.
The speed is given to the nearest 10 m.p.h. and the time to the nearest hour.
Calculate the shortest and longest possible distances travelled by the aircraft.

c Subtracting
d Dividing

1 Parminder planted a tree. Its height was 1.7 m, to the nearest 0.1 m.
A year later the height of the tree was 2.7 m, to the nearest 0.1 m.

   (a) Write down the upper and lower bounds for the height of the tree when planted.

   (b) Write down the upper and lower bounds for the height of the tree a year later.

   (c) Find the upper and lower bounds for the tree's gain in height over the year.

2 A piece of wood is 1.6 m long, correct to the nearest 10 cm.
Jack cuts off a piece 72 cm long, correct to the nearest centimetre.
Find the upper and lower bounds of the length of wood remaining.

3 A bottle of medicine contains 100 ml, correct to the nearest millilitre.
One dose of medicine is 5 ml.
If the doses are measured correct to the nearest millilitre, what is
the minimum number of doses in the bottle?

4 A pile of coal weighs 376 tonnes, to the nearest tonne.
10 truckloads are removed, each being 25 tonnes to the nearest tonne.
Find the greatest and least possible values of the weight of coal left in the pile.

5 The power dissipated in an electrical circuit is given by the formula $P = \dfrac{V^2}{R}$

where P is the power in watts, V is the voltage and R is the resistance in ohms.
Calculate the upper and lower bounds of the power if the voltage is 240 volts and
the resistance is 410 ohms, both measured to two significant figures.

6 In this right-angled triangle, side $a = 4.7$ cm
and side $c = 5.4$ cm, both to the nearest 0.1 cm.

Calculate the upper and lower bounds for the side b.
Give each answer to four significant figures.

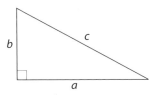

7 The table shows the population in 2006
and the area of the countries of the UK,
to three significant figures.

Calculate the upper and lower bounds
for the population density
(people per km²) of each country in 2006.
Give your answers correct to four significant figures.

Country	Population	Area (km²)
England	5.08×10^7	1.30×10^5
Wales	2.97×10^6	2.07×10^4
Scotland	5.12×10^6	7.79×10^4
Northern Ireland	1.74×10^6	1.36×10^4

22 Using exact values

A Showing π in a result

1 Which expression below gives

   (a) the exact area of this circle in square centimetres

   (b) the exact circumference of this circle in centimetres

| 3π | 6π | 9π | 12π | 18π | 36π |

2 The shapes below are drawn on a grid of centimetre squares.
Find the perimeter and area of each shape in terms of π.

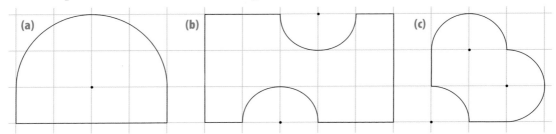

  (a)   **(b)**   **(c)**

3 This diagram shows a sector of a circle.

   (a) Find the exact area of the sector.

   (b) Find the exact perimeter of the sector.

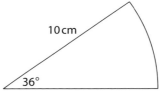

B Showing a surd in a result

1 These shapes are drawn on a grid of centimetre squares.
Find the perimeter of each shape in surd form.

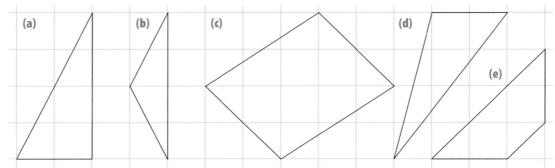

  (a)   **(b)**   **(c)**   **(d)**   **(e)**

2 Find the exact perimeter of this triangle.

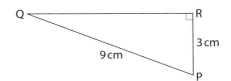

3 Find an exact expression for

(a) AC

(b) AD

(c) the area of ABCD

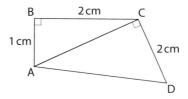

4 Given that $\tan x = \frac{2}{5}$, write exact values for $\sin x$ and $\cos x$.

*5 ABC is an isosceles right-angled triangle.

AD bisects the angle of 45° at A, making two angles of $22\frac{1}{2}°$.
AD is the locus of points equidistant from lines AB and AC,
so it follows that DB = DE, where DE is perpendicular to AC.

(a) What can you say about triangles ABD and AED?

(b) What is the length of AE?

(c) Write the length AC in surd form.

(d) Write the length EC in surd form.

(e) Explain why EC = BD.

(f) Use triangle ABD to find the value of $\tan 22\frac{1}{2}°$ in surd form.

(g) On a calculator find $\tan 22\frac{1}{2}°$ and check that this agrees with your answer to (f).

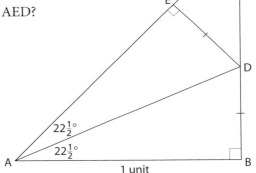

ᴄ **Mixed questions**

1 A cylinder has radius 4 cm and height 5 cm.
Find in terms of π

(a) the volume of the cylinder

(b) the total surface area of the cylinder

2 The diagram shows a running track.
Find in terms of π

(a) the difference in length between the outer and inner edges of the track

(b) the area of the track (shaded)

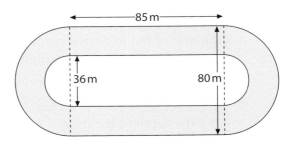

3 These shapes are drawn on a centimetre squared grid.
Find their perimeters and areas as exact values.

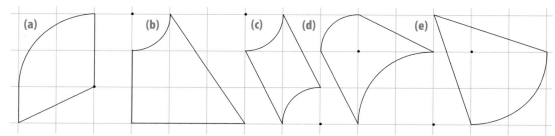

*4 A rectangle of sides a and b is drawn inside a circle.
A semicircle is drawn on each of the four sides of
the rectangle, as shown here.

(a) Explain why the area of the white circle is

$\frac{1}{4}\pi(a^2 + b^2)$

(b) Show that the shaded area is equal to the area
of the rectangle.

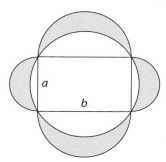

D Exact relationships between variables

1 Two circles the same size fit exactly inside a larger circle as shown.
What exact fraction of the area of the larger circle is
the area shown shaded?

2 A cone and a sphere have the same volume.
The sphere and the base of the cone have the same radius, r.
Express the the height h of the cone in terms of r.

3 A circle of radius a has the same area as the sector shown here.
Find a as an exact value.

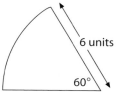

6 units

60°

4 A semicircle of radius 6 cm is made into the curved
surface of a cone (without overlapping).

(a) What is the radius of the 'base' of the cone?

(b) Find an expression for the exact value of

(i) the depth of the cone

(ii) the volume of the cone

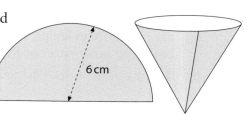

6 cm

23 Fractional indices

1 Write each of these as a single power of 3.

   (a) 81   (b) 81^3 (c) $\frac{9^5}{3}$ (d) $\frac{1}{27}$ (e) $\left(\frac{1}{9}\right)^4$

2 Find the value of n in each of these equations.

   (a) $3^n = 1$ (b) $10^n = \frac{1}{10}$ (c) $2^n = \frac{1}{32}$ (d) $\left(\frac{1}{2}\right)^n = 16$

3 Evaluate each of these as an integer or fraction.

   (a) $\sqrt[3]{125}$ (b) $\sqrt[3]{\frac{27}{64}}$ (c) $\sqrt[4]{81}$ (d) $\sqrt[6]{\frac{1}{64}}$ (e) $\sqrt[8]{1}$

C Fractional indices of the form 1/n or ⁻1/n
D Fractional indices of the form p/q or ⁻p/q

1 Find the value of each of these as an integer or fraction.

   (a) $144^{\frac{1}{2}}$ (b) $\left(\frac{1}{9}\right)^{\frac{1}{2}}$ (c) $8^{\frac{1}{3}}$ (d) $10\,000^{\frac{1}{4}}$ (e) $1^{\frac{1}{5}}$

   (f) $\left(\frac{9}{16}\right)^{\frac{1}{2}}$ (g) $25^{-\frac{1}{2}}$ (h) $64^{-\frac{1}{3}}$ (i) $\left(\frac{1}{16}\right)^{-\frac{1}{4}}$ (j) $\left(\frac{27}{125}\right)^{-\frac{1}{3}}$

2 Find the value of each of these as an integer or fraction.

   (a) $64^{\frac{2}{3}}$ (b) $81^{\frac{3}{4}}$ (c) $32^{\frac{2}{5}}$ (d) $4^{\frac{3}{2}}$ (e) $\left(\frac{8}{27}\right)^{\frac{2}{3}}$

   (f) $\left(\frac{16}{81}\right)^{\frac{3}{4}}$ (g) $8^{-\frac{2}{3}}$ (h) $81^{-\frac{3}{4}}$ (i) $\left(\frac{1}{27}\right)^{-\frac{2}{3}}$ (j) $\left(\frac{9}{25}\right)^{-\frac{3}{2}}$

3 Simplify each of these as an integer or fraction.

   (a) $121^{\frac{1}{2}} \times 16^{\frac{1}{2}}$ (b) $5^{-2} \times 25^{\frac{1}{2}}$ (c) $2^{-3} \times 32^{\frac{4}{5}}$ (d) $\left(\frac{27}{125}\right)^{-\frac{2}{3}} \times \left(\frac{4}{25}\right)^{\frac{1}{2}}$

4 Write each of these as a single power of 2.

   (a) $\left(\sqrt{2}\right)^7$ (b) $16\sqrt{2}$ (c) $2 \div \sqrt[3]{2}$ (d) $4 \times \sqrt{32}$

5 Find the value of n in each statement.

   (a) $100^n = 10$ (b) $81^n = \frac{1}{9}$ (c) $8^n = 16^{\frac{1}{2}}$ (d) $n \times \sqrt{8} = 2^{\frac{7}{2}}$

E Powers and roots on a calculator

1 Use a calculator to find the following, to four significant figures.

   (a) $5^{0.5}$ (b) $\sqrt[4]{8}$ (c) $6.5^{-1.8}$ (d) $0.9^{\frac{2}{3}}$ (e) $\sqrt[5]{3^8}$

24 Exponential growth and decay

You need graph paper.

A Exponential growth
B Exponential decay

1 The population of a colony of animals increases by the same percentage each year.
Its population P is given by the equation $P = 100 \times 1.3^t$,
where t is the time in years since the first measurement.

(a) What was the population of the colony when first measured?

(b) By what percentage does the population increase each year?

(c) Sketch the graph of P against t.

(d) Calculate, to two significant figures, the value of P when $t = 2.5$.

(e) Calculate, to two significant figures, the value of P when $t = {}^-2$.
What does this result tell you?

2 An object is heated and then allowed to cool.
Its temperature $T\,°C$ after cooling for t minutes is given by the formula $T = 45 \times 1.2^{-t}$.

(a) What was the temperature of the object when cooling began?

(b) Calculate, to the nearest °C, its temperature after cooling for 8 minutes.

(c) Draw a sketch to show the shape of the graph of T against t.

3 The population of a region appears to fit the formula $P = 5000 \times 2^{0.01n}$,
where P is the population n years after the year 2000.

(a) What was the population in 2000?

(b) Calculate, to two significant figures, the population in 2020.

(c) Show that the population is expected to double by 2100.

4 The population of a town is 42 600 now. The population decreases by 2% every year.

(a) Calculate, to three significant figures, the population 5 years from now.

(b) If P is the population after t years, write down an equation connecting P and t.

(c) Assuming that the population has been decreasing at the same rate in the past,
calculate the population 10 years ago.

5 Martha invests a sum of money in an account that pays interest of 5.5% per annum.
Show that her investment will double in value after 13 years.

6 (a) On graph paper, draw the graph of $y = 2^{-x}$ for values of x from ${}^-3$ to 3.

(b) Use your graph to estimate the solution to the equation $2^{-x} = 5$.

25 Angle properties of a circle

B Angles in a circle

1 Find the angles marked with letters.
(The centre of each circle is marked with a dot.)

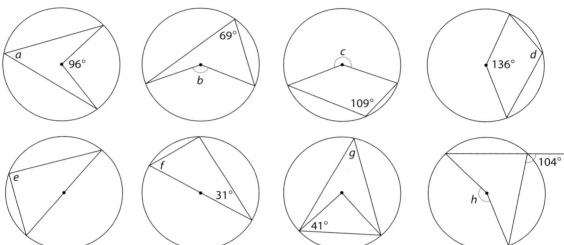

2 Calculate the angles marked with small letters,
giving a reason for each step of working.

(a) **(b)** **(c)**

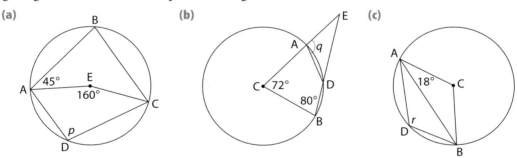

3 Find the angles marked with letters.

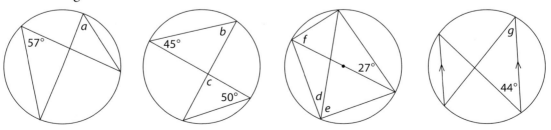

4 Calculate the angles marked with small letters, giving a reason for each step of working.

(a)

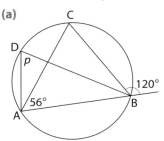

(b)

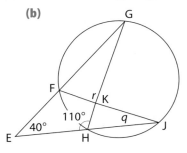

(c)

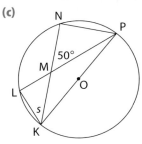

C Angles of a cyclic quadrilateral

1 Find the angles marked with letters.

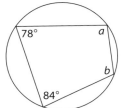
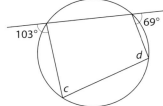
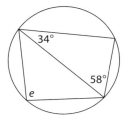
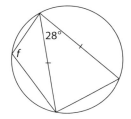

2 Calculate the angles marked with small letters, giving a reason for each step of working.

(a)

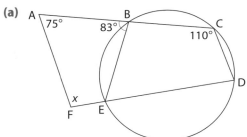

(b)

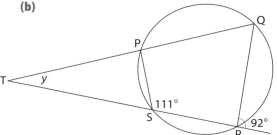

D Tangent to a circle

1 Each of these diagrams includes a tangent to a circle.
Find the angles marked with letters.

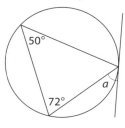
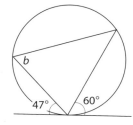
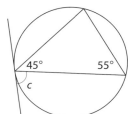
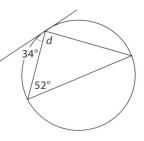

2 These diagrams include tangents.
Find the angles marked with small letters, giving full reasons.

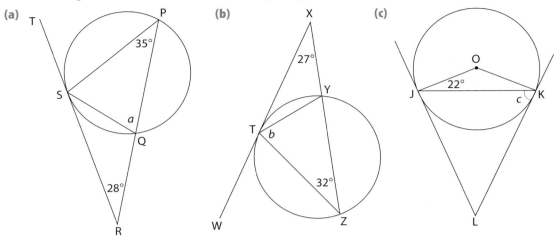

(a)

(b)

(c)

1 Find the angles marked with letters.

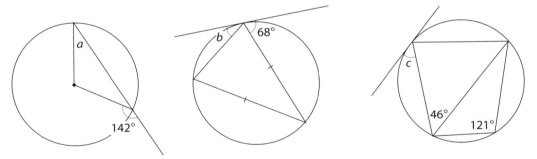

2 Find the angles marked with small letters, giving full reasons.

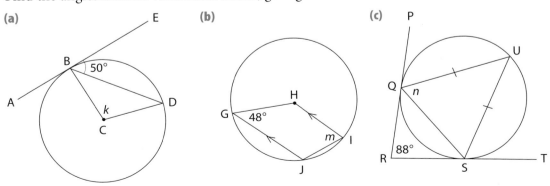

(a)

(b)

(c)

26 Algebraic fractions and equations 2

A Adding and subtracting algebraic fractions

1 Write each of these as a single fraction in its simplest form.

(a) $\dfrac{a}{6} + \dfrac{a}{3}$　　(b) $\dfrac{a}{4} + \dfrac{a}{3}$　　(c) $\dfrac{4a}{3} - \dfrac{2a}{5}$　　(d) $\dfrac{3a}{4} - \dfrac{5a}{12}$

(e) $\dfrac{a}{12} + \dfrac{b}{6}$　　(f) $\dfrac{p}{4} - \dfrac{q}{5}$　　(g) $\dfrac{x}{20} + \dfrac{3y}{5}$　　(h) $\dfrac{3m}{5} - \dfrac{5n}{3}$

2 Write each of these as a single fraction in its simplest form.

(a) $\dfrac{a+1}{2} - \dfrac{a}{3}$　　(b) $\dfrac{2a+3}{6} + \dfrac{a}{3}$　　(c) $\dfrac{3a+5}{2} + \dfrac{a}{10}$　　(d) $\dfrac{2a}{3} - \dfrac{a+3}{4}$

(e) $1 + \dfrac{2a}{5}$　　(f) $2 + \dfrac{3a+1}{4}$　　(g) $6 - \dfrac{2a+3}{5}$　　(h) $7 - \dfrac{3a-2}{4}$

(i) $\dfrac{a+3}{2} + \dfrac{2a-1}{3}$　　(j) $\dfrac{6a+3}{4} + \dfrac{3a+1}{5}$　　(k) $\dfrac{a+5}{2} - \dfrac{a+3}{4}$　　(l) $\dfrac{2a-5}{4} - \dfrac{a-2}{6}$

3 Write each of these as a single fraction.

(a) $\dfrac{2}{3a} + \dfrac{1}{2a}$　　(b) $\dfrac{1}{3a} - \dfrac{1}{a^2}$　　(c) $\dfrac{5}{a} + \dfrac{a}{2}$　　(d) $\dfrac{6a}{5} - \dfrac{1}{3a}$

(e) $\dfrac{1}{a} - \dfrac{1}{b}$　　(f) $\dfrac{2}{x} + \dfrac{1}{5y}$　　(g) $\dfrac{1}{2k} + \dfrac{2}{3h}$　　(h) $\dfrac{3a+7}{a} - \dfrac{a+6}{2}$

B More complex denominators

1 Write each of these as a single fraction in its simplest form.

(a) $\dfrac{1}{x} + \dfrac{1}{x+3}$　　(b) $\dfrac{2}{x+4} + \dfrac{1}{2x-7}$　　(c) $\dfrac{1}{x+3} - \dfrac{1}{x+6}$　　(d) $\dfrac{x}{3x+1} + \dfrac{2}{x+3}$

(e) $x + \dfrac{1}{x-3}$　　(f) $\dfrac{x}{x+1} - \dfrac{7}{x+7}$　　(g) $\dfrac{5}{x-3} - 2x$　　(h) $\dfrac{3x}{x-5} - \dfrac{2x}{x+1}$

2 Prove that $\dfrac{2}{x+5} + \dfrac{3}{x-5} = \dfrac{5(x+1)}{x^2-25}$.

3 Write each of these as a single fraction in its simplest form.

(a) $\dfrac{x+3}{x+1} + \dfrac{1}{x}$　　(b) $\dfrac{x-2}{x} - \dfrac{x}{3x-2}$　　(c) $\dfrac{x+1}{x-4} - \dfrac{1}{x+4}$　　(d) $\dfrac{2x}{x+1} - \dfrac{x-1}{x+7}$

4 Write each of these as a single fraction in its simplest form.

(a) $\dfrac{1}{(x+3)^2} + \dfrac{2}{x+3}$　　(b) $\dfrac{x}{x^2-1} - \dfrac{3}{x+1}$　　(c) $\dfrac{x}{5x-25} - \dfrac{x}{x^2-25}$

C Solving equations containing fractions

1 Solve each of these equations.
  Give all solutions as integers or fractions.

  (a) $\dfrac{x+1}{3} + \dfrac{x-2}{5} = 9$

  (b) $\dfrac{3x-1}{2} - \dfrac{x}{5} = x$

  (c) $\dfrac{2x+1}{3} + \dfrac{5x}{4} = 2x - 1$

  (d) $\dfrac{x+5}{2} + \dfrac{x-1}{3} = x + 4$

  (e) $\dfrac{x+1}{4} - \dfrac{x-3}{2} = x + 13$

  (f) $\dfrac{x+7}{5} - \dfrac{2x-1}{3} = 3x$

2 Solve each of these equations.
  Give all solutions as integers or fractions.

  (a) $\dfrac{5}{x} + \dfrac{2}{3x} = 17$

  (b) $\dfrac{x}{x+2} + \dfrac{4}{x+7} = 1$

  (c) $\dfrac{3x}{x+1} + \dfrac{x}{x-1} = 4$

  (d) $\dfrac{x-1}{x+1} = \dfrac{x+2}{x+6}$

  (e) $\dfrac{2x}{x-2} + \dfrac{1}{x+1} = 2$

  (f) $\dfrac{3x}{x-2} - \dfrac{4}{x} = 3$

  (g) $\dfrac{8}{x} + \dfrac{3x}{x+2} = 4$

  (h) $\dfrac{3}{x} + 4x = 13$

  (i) $\dfrac{x}{5x-12} = \dfrac{2}{x}$

  (j) $\dfrac{5}{2x+1} + \dfrac{6}{x+1} = 3$

  (k) $\dfrac{1}{x-3} - \dfrac{3}{x+2} = \dfrac{1}{2}$

  (l) $\dfrac{9}{x+1} - \dfrac{4x}{x+2} = 1$

3 Solve each of these equations.
  Give all solutions as decimals correct to 2 d.p.

  (a) $\dfrac{3x}{3x+1} - \dfrac{1}{x-5} = 2$

  (b) $\dfrac{x+2}{x-1} = \dfrac{2x-5}{x+3}$

  (c) $\dfrac{x+5}{x-2} - \dfrac{3}{5-x} = 5$

4 One day Pam cycles for x hours and travels a distance of $112\,\text{km}$.

  (a) Write down, in terms of x, Pam's average speed in km/h.

  The next day she takes one hour longer to cycle the same distance.
  Her average speed is $2\,\text{km/h}$ slower than the day before.

  (b) Show that $x^2 + x - 56 = 0$.

  (c) Find the number of hours Pam cycles on the second day.

D Rearranging formulas containing fractions

1 Rearrange each formula to make y the subject.

  (a) $P = \dfrac{y}{5} - \dfrac{x}{z}$

  (b) $z = \dfrac{1}{y} + \dfrac{1}{x}$

  (c) $\dfrac{2}{5x} + \dfrac{z}{2y} = 1$

  (d) $\dfrac{1}{z} = \dfrac{1}{x} + \dfrac{1}{y}$

  (e) $N = \dfrac{2}{3y} - \dfrac{1}{x}$

  (f) $k = \dfrac{1}{3m} - \dfrac{1}{y^2}$

27 Vectors

You need squared paper.

A Vector notation

1 This diagram shows vectors **a** and **b**.
On squared paper draw and label vectors equivalent to

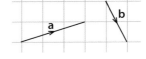

   (a) 2**a**   (b) 5**b**   (c) ⁻**a**

   (d) ⁻**b**   (e) ⁻3**a**   (f) ⁻4**b**

2 $\mathbf{c} = \begin{bmatrix} 5 \\ 2 \end{bmatrix}$ and $\mathbf{d} = \begin{bmatrix} 4 \\ -1 \end{bmatrix}$.

   (a) Write each of these as a column vector.

      (i) 3**c**   (ii) 2**d**   (iii) ⁻**d**   (iv) ⁻2**c**

   (b) Write each of these in terms of **c** or **d**.

      (i) $\begin{bmatrix} 10 \\ 4 \end{bmatrix}$ (ii) $\begin{bmatrix} 12 \\ -3 \end{bmatrix}$ (iii) $\begin{bmatrix} -5 \\ -2 \end{bmatrix}$ (iv) $\begin{bmatrix} -16 \\ 4 \end{bmatrix}$

B Adding vectors
C Subtracting vectors

1 $\mathbf{a} = \begin{bmatrix} 2 \\ 3 \end{bmatrix}$, $\mathbf{b} = \begin{bmatrix} 1 \\ -2 \end{bmatrix}$ and $\mathbf{c} = \begin{bmatrix} -4 \\ 1 \end{bmatrix}$. What single column vector is equivalent to

   (a) **a** + **b**   (b) **a** + **c**   (c) **b** + 2**c**   (d) 4**b** + 3**a**   (e) 3**a** + 2**c**

2 PQRS is a parallelogram. $\overrightarrow{PQ} = \mathbf{x}$ and $\overrightarrow{PS} = \mathbf{y}$.
Express each of these vectors in terms of **x** and **y**.

   (a) $\overrightarrow{SR}$ (b) $\overrightarrow{RQ}$ (c) $\overrightarrow{PR}$

   (d) $\overrightarrow{SQ}$ (e) $\overrightarrow{QS}$

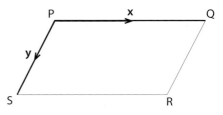

3 This is part of a grid of congruent parallelograms.
$\overrightarrow{OA} = \mathbf{a}$ and $\overrightarrow{OB} = \mathbf{b}$.
Write these vectors in terms of **a** and **b**.

   (a) $\overrightarrow{OF}$ (b) $\overrightarrow{OD}$ (c) $\overrightarrow{OG}$

   (d) $\overrightarrow{AK}$ (e) $\overrightarrow{JG}$ (f) $\overrightarrow{ED}$

   (g) $\overrightarrow{AI}$ (h) $\overrightarrow{HC}$ (i) $\overrightarrow{LF}$

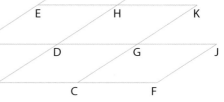

4 ABCDEFGH is a regular octagon.

$\overrightarrow{OA} = \mathbf{u}$ $\qquad$ $\overrightarrow{OB} = \mathbf{v}$ $\qquad$ $\overrightarrow{OC} = \mathbf{w}$

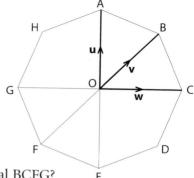

(a) Express these in terms of **u**, **v** and **w**.

(i) $\overrightarrow{GC}$ $\qquad$ (ii) $\overrightarrow{OF}$

(iii) $\overrightarrow{AB}$ $\qquad$ (iv) $\overrightarrow{AC}$

(b) Express these in terms of **v** and **w**.

(i) $\overrightarrow{BC}$ $\qquad$ (ii) $\overrightarrow{GB}$

(iii) $\overrightarrow{GF}$ $\qquad$ (iv) $\overrightarrow{FC}$

(c) What do your results in (b) tell you about quadrilateral BCFG?

5 A, B and C are three points such that $\overrightarrow{BC} = 2\overrightarrow{AB}$.
What can you say about the three points?

D **Fractions of a vector**
E **Vector algebra**

1 This diagram shows vectors **p** and **q**.

On squared paper draw and label vectors

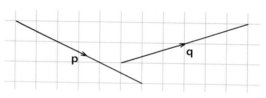

(a) $\frac{1}{2}\mathbf{q}$ $\qquad$ (b) $\frac{1}{3}\mathbf{p}$ $\qquad$ (c) $\frac{5}{6}\mathbf{p}$

(d) $1\frac{2}{3}\mathbf{p}$ $\qquad$ (e) $\frac{-2}{3}\mathbf{p}$ $\qquad$ (f) $\frac{-3}{2}\mathbf{q}$

2 Vectors **x** and **y** are shown on a grid of congruent parallelograms.
$\overrightarrow{OX} = \mathbf{x}$ and $\overrightarrow{OY} = \mathbf{y}$.

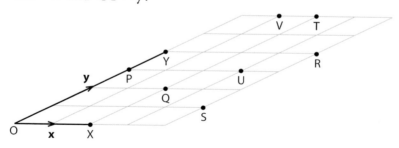

Write these vectors in terms of **x** and **y**.

(a) $\overrightarrow{OP}$ $\qquad$ (b) $\overrightarrow{VT}$ $\qquad$ (c) $\overrightarrow{OQ}$ $\qquad$ (d) $\overrightarrow{RS}$ $\qquad$ (e) $\overrightarrow{QU}$

(f) $\overrightarrow{SU}$ $\qquad$ (g) $\overrightarrow{TU}$ $\qquad$ (h) $\overrightarrow{RV}$ $\qquad$ (i) $\overrightarrow{VS}$

3 Write each vector expression in a simpler form.

(a) $\mathbf{x} + \mathbf{y} + \mathbf{x} + 2\mathbf{y}$ $\qquad$ (b) $2\mathbf{x} + \mathbf{y} - 4\mathbf{y}$ $\qquad$ (c) $2\mathbf{x} + 5(\mathbf{x} + 2\mathbf{y})$

(d) $3(\mathbf{x} - 2\mathbf{y}) + 6\mathbf{y}$ $\qquad$ (e) $\frac{1}{2}\mathbf{x} + \frac{3}{4}\mathbf{y} + \frac{1}{2}\mathbf{x} - \frac{1}{4}\mathbf{y}$ $\qquad$ (f) $\frac{1}{4}\mathbf{x} + \frac{2}{3}\mathbf{y} - \frac{1}{2}\mathbf{x} - \frac{1}{2}\mathbf{y}$

(g) $\mathbf{y} + 4(\mathbf{x} - 3\mathbf{y})$ $\qquad$ (h) $\frac{1}{2}\mathbf{y} + \frac{1}{4}(\mathbf{x} + 2\mathbf{y})$ $\qquad$ (i) $\frac{2}{3}\mathbf{y} + \frac{1}{2}(\mathbf{x} - \mathbf{y})$

4 This shows points O, M and N and
the vectors **m** and **n**.

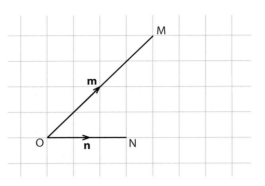

Copy the diagram.
Mark points P, Q, R, S, T and U to
give these vectors.

(a) $\overrightarrow{OP} = \mathbf{n} + \frac{1}{2}\mathbf{m}$ (b) $\overrightarrow{OQ} = \mathbf{m} + \frac{1}{3}\mathbf{n}$

(c) $\overrightarrow{OR} = 2\mathbf{n} + \frac{1}{4}\mathbf{m}$ (d) $\overrightarrow{OS} = \frac{1}{2}\mathbf{m} + \frac{2}{3}\mathbf{n}$

(e) $\overrightarrow{OT} = \frac{3}{4}\mathbf{m} - \mathbf{n}$ (f) $\overrightarrow{OU} = \mathbf{n} - \frac{1}{4}\mathbf{m}$

5 ABCD is a trapezium.
$\overrightarrow{DC} = 3\overrightarrow{AB}$

(a) Write these as simply as possible in terms of **u** and **v**.

(i) $\overrightarrow{DC}$ (ii) $\overrightarrow{AC}$ (iii) $\overrightarrow{DB}$ (iv) $\overrightarrow{AD}$

(b) X is the point on AD for which $\overrightarrow{AX} = \frac{1}{3}\overrightarrow{AD}$.
Write $\overrightarrow{XC}$ as simply as possible in terms of **u** and **v**.

6 In triangle OAB, point M is the mid-point of AB.
Write $\overrightarrow{OM}$ as simply as possible in terms of **a** and **b**.

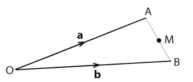

7 ABCDEF is a regular hexagon.
Write $\overrightarrow{AF}$ as simply as possible in terms of **x** and **y**.

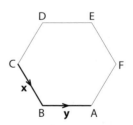

F **Proof by vectors**

1 ABCD is a parallelogram.
P is the point on AB such that AP = 2PB.
Q is the point on DB such that DQ = 3QB.
$\overrightarrow{DA} = \mathbf{x}$ and $\overrightarrow{DC} = \mathbf{y}$.

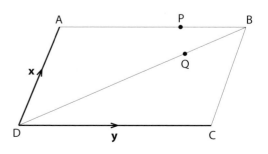

(a) Find, in terms of **x** and **y**, expressions for

(i) $\overrightarrow{PQ}$ (ii) $\overrightarrow{PC}$

(b) Show that PQC is a straight line.

2 OABC is a parallelogram.

$\overrightarrow{OA} = \mathbf{a}$ $\qquad$ $\overrightarrow{OC} = \mathbf{c}$

C is the point on BE such that $\overrightarrow{BC} = \frac{1}{4}\overrightarrow{BE}$.

D is the point on OC such that $\overrightarrow{OD} = \frac{1}{3}\overrightarrow{OC}$.

(a) Express $\overrightarrow{OE}$ in terms of **a** and **c**.

(b) **(i)** Prove that AD and OE are parallel.

$\qquad$ **(ii)** What is the ratio of lengths AD and OE?

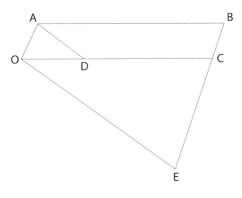

3 OACB is a parallelogram.

$\overrightarrow{OA} = \mathbf{a}$ $\qquad$ $\overrightarrow{OB} = \mathbf{b}$

E is the mid-point of OA.

F is the point on OB such that $\overrightarrow{OF} = \frac{1}{4}\overrightarrow{OB}$.

The point G divides BC in the ratio $3:5$.

(a) Find, in terms of **a** and **b**,

an expression for $\overrightarrow{FG}$.

(b) Show that $\overrightarrow{FG} = \frac{3}{4}\overrightarrow{EC}$.

(c) What type of quadrilateral is EFGC?
Justify your answer

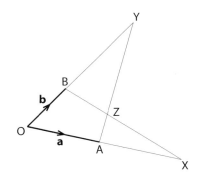

*4 In the diagram $\overrightarrow{OA} = \mathbf{a}$ and $\overrightarrow{OB} = \mathbf{b}$.

A is the mid-point of OX.

B divides OY in the ratio $1:2$.

(a) Write expressions, in terms of **a** and **b**, for

$\qquad$ **(i)** $\overrightarrow{OY}$ $\qquad\qquad$ **(ii)** $\overrightarrow{AY}$

$\qquad$ **(iii)** $\overrightarrow{OX}$ $\qquad\qquad$ **(iv)** $\overrightarrow{BX}$

(b) $\overrightarrow{AZ} = h\overrightarrow{AY}$ where h is a fraction.

Write and simplify an expression for $\overrightarrow{OZ}$
in terms of **a**, **b** and h.

(c) Similarly, $\overrightarrow{BZ} = k\overrightarrow{BX}$.

Write an expression for $\overrightarrow{OZ}$ in terms of **a**, **b** and k.

(d) Since the two expressions for $\overrightarrow{OZ}$ found in (b) and (c) must be equal,
form a pair of simultaneous equations in h and k.
Use these to find the value of h.

(e) Hence write down the ratio in which Z divides AY.

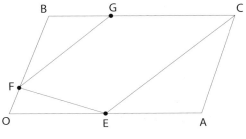

Mixed practice 4

1 Share £81 in the ratio $4:3:2$.

2 The pie chart shows information about the number of
loaves of bread sold in a shop one day.

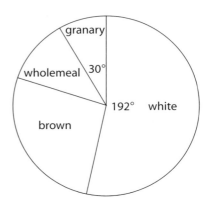

    (a) Which type of bread is the mode?

    (b) 25 loaves of granary bread were sold.
How many loaves of bread were sold altogether?

    (c) What fraction of the loaves sold were white?
Write this fraction in its simplest form.

    (d) 80 brown loaves were sold.
What is the angle for brown loaves?

    (e) How many wholemeal loaves were sold that day?

3 How long, in hours and minutes, does it take a train to travel 200 miles at 75 m.p.h.?

4 The price of a washing machine is reduced by 15% in a sale to £331.50.
Calculate the original price of the washing machine.

5 This net is made from four congruent triangles
and a square.

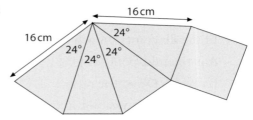

    (a) What three-dimensional shape will
this net make when folded up?

    (b) Calculate the total surface area
of the shape, correct to 3 s.f.

6 Marji opens a building society account with £200.
The building society pays compound interest at a rate of 5.5% a year.
How many years will it take for the money in the account to grow to at least £250?

7 The tangents at P and Q to the circle centre O meet at R.

Find angles a, b and c, giving reasons for your answers.

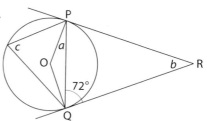

8 The formula connecting a, b and c is given as $c = \dfrac{2a}{2 - ab}$.

Given that $a = \frac{5}{7}$ and $b = \frac{2}{5}$, find the value of c as a fraction in its simplest form.

9 Evaluate these.

    (a) $\sqrt[3]{\frac{1}{27}}$ **(b)** $16^{\frac{1}{2}}$ **(c)** $8000^{\frac{1}{3}}$ **(d)** $8^{\frac{2}{3}}$ **(e)** $\left(\frac{1}{9}\right)^{-\frac{1}{2}}$

10 In this diagram P, Q, R and S are points on the circumference of a circle centre O.

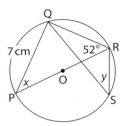

(a) Find the sizes of angles x and y, giving reasons for your answers.

(b) Calculate the length of the radius, correct to 3 s.f.

11 The centre of the square PQRS is at O.
T, U, V, W are the mid-points of the sides of the square.
$\overrightarrow{OP} = \mathbf{p}$ and $\overrightarrow{OT} = \mathbf{t}$.

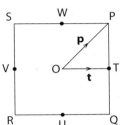

Express each of these vectors in terms of $\mathbf{p}$ and $\mathbf{t}$.

(a) $\overrightarrow{TP}$　　(b) $\overrightarrow{VP}$　　(c) $\overrightarrow{OQ}$　　(d) $\overrightarrow{US}$

(e) $\overrightarrow{RT}$　　(f) $\overrightarrow{SQ}$　　(g) $\overrightarrow{WR}$　　(h) $\overrightarrow{PU}$

12 The surface of a drive is to be a 10 cm layer of tarmac, measured to the nearest 1 cm. The drive has been measured as 9 m long and 3.6 m wide, both to the nearest 10 cm. Calculate the upper bound for the volume of tarmac required for the drive.

13 Solve these equations.

(a) $\dfrac{2}{x} - \dfrac{1}{x-1} = 6$　　(b) $\dfrac{x}{x-3} + \dfrac{14}{x+1} = 4$　　(c) $\dfrac{3}{x} - \dfrac{x-3}{2x+7} = 5$

14 A, B, C and D are points on a circle.
AB and CD meet at point X.

$AX = 4\,cm$, $XB = 3\,cm$ and $CX = 5\,cm$.

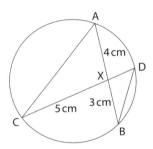

(a) Explain why triangles AXC and DXB are similar.

(b) Calculate XD.

15 Hayley cycles for 8 km at x km/h and then runs for 6 km at $(x-10)$ km/h. The total time she spends cycling and running is 1 hour.

(a) Form an equation in x and show that it simplifies to $x^2 - 24x + 80 = 0$.

(b) Solve this equation and find the speeds at which Hayley runs and cycles.

16 A, B, C and D are points on the circumference of a circle. O is the centre of the circle.

Calculate these angles, giving reasons for each step of your working.

(a) Angle ABC

(b) Angle BCO

(c) Angle ADC

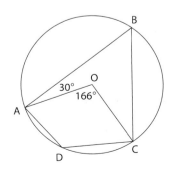

17 In this diagram, $\overrightarrow{OA} = \mathbf{a}$ and $\overrightarrow{OB} = \mathbf{b}$.

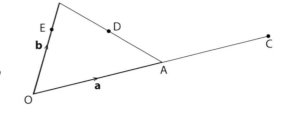

C is the point such that $\overrightarrow{AC} = \mathbf{a}$.
D is the mid-point of AB.
E is two-thirds of the way along OB.

(a) Express $\overrightarrow{AB}$ and $\overrightarrow{AD}$ in terms of $\mathbf{a}$ and $\mathbf{b}$

(b) Prove that EDC is a straight line.

(c) Write down the ratio ED : DC.

18 Calculate the exact length of the line segment joining $(^-3, 5)$ to $(7, 1)$.

19 Emma makes her own Christmas cards and uses a star and a bow on each one.
Stars are sold in packs of 30. Bows are sold in packs of 18.
She buys some packs of each and uses them all.
What is the least number of cards that Emma makes?

20 A, B, C and D are four points on the circumference of a circle.
BD is a diameter of the circle and DE is a tangent to the circle at D.
Angle DAC = 30°.

Calculate these angles, giving reasons for each step of your working.

(a) Angle BED   (b) Angle CDE

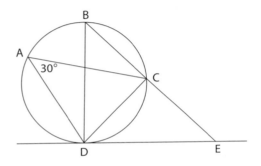

21 Find the value of $4^{-\frac{1}{2}} + 4^0 + 4^{\frac{1}{2}} + 4^{\frac{3}{2}}$.

22 The cross-section of a solid is the sector of a circle with radius 6 cm and angle 120°.
The height of the solid is 3 cm.

(a) Calculate the volume of the solid. Give your answer in terms of π.

(b) Show that the surface area of the solid is $36(\pi + 1)\,\text{cm}^2$.

23 Pritti weighed a potato and then dried it in an oven.
Before drying it weighed 84 g and afterwards 55 g, both correct to the nearest gram.
Calculate, to the nearest 0.1%, the upper and lower bounds of the percentage weight loss.

24 The population P of a town t years after the start of 2006 is given by $P = 152\,000 \times 0.98^t$.

(a) What was the population of the town at the start of 2006?

(b) Calculate, correct to three significant figures, the population at the start of 2010.

25 Sam is doing a cycling time trial.
He wants to cover the 20 km distance at an average speed of 40 km/h.
The first 5 km is uphill and he only averages 15 km/h.
To reach his target, what would his average speed need to be for the rest of the trial?

28 Circles and equations

You need graph paper for section B.

A Equation of a circle

1 Sketch the circle with equation $x^2 + y^2 = 81$.

2 Show that $(^-4, 5)$ is a point on the circle with equation $x^2 + y^2 = 41$.

3 Write down the equation of each circle.

(a)

(b)
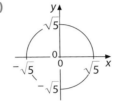

(c)
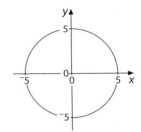

4 Write down the exact value of the radius of each circle.

    (a) $x^2 + y^2 = 144$ **(b)** $x^2 + y^2 = 49$ **(c)** $x^2 + y^2 = 30$

5 Write down the equations of the circles with centre $(0, 0)$ and the following radii.

    (a) 11   **(b)** 14   **(c)** $\frac{1}{5}$ **(d)** $\frac{2}{3}$ **(e)** $\sqrt{12}$

6 A circle has centre at the origin and passes through point $(0, 8)$.

    (a) What is its equation?

    (b) Give the coordinates of the points where it meets the x-axis.

7 A circle has centre at the origin and passes through the point $(6, 8)$.
Write down the equation of the circle.

B Intersection of a line and a circle

1 The diagram shows a sketch of the circle
$x^2 + y^2 = 40$ and the line $y = x + 4$.

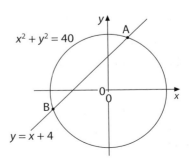

    (a) For the points of intersection A and B
show that $x^2 + 4x - 12 = 0$.

    (b) Hence find the coordinates of the points
of intersection of the line $y = x + 4$
and the circle $x^2 + y^2 = 40$.

2 Find the coordinates of the two points where
the line $x + y = 2$ crosses the circle $x^2 + y^2 = 52$.

3 Find the points of intersection of the circle $x^2 + y^2 = 65$
and the line $y = 2x - 10$.

4 Use graph paper for this question, using a scale of 2 cm to one unit,
with both axes numbered from $^-4$ to 4.

(a) Draw the line $y = x - 2$.

(b) Draw the circle with centre at the origin and radius 4.

(c) Use the graph to write down the coordinates of the points where
the line crosses the circle, as accurately as you can.

(d) Write down the equation of the circle.

(e) By solving a pair of simultaneous equations, write down
the coordinates of the points of intersection of the line and the circle,
correct to two decimal places.

5 (a) (i) Without sketching, show that the point $(1, 1)$ is inside the circle $x^2 + y^2 = 7$.

(ii) Hence show that the line $y = 2x - 1$ cuts the circle at two points.

(b) Find, correct to two decimal places, the coordinates of the points of intersection
of the line $y = 2x - 1$ and the circle $x^2 + y^2 = 7$.

6 Match up the following three pairs of equations with the three statements below.
Explain your method.

(a) $x^2 + y^2 = 14$
$y = 3x - 8$

(b) $x^2 + y^2 = 6$
$y = x + 4$

(c) $x^2 + y^2 = 18$
$y = x + 6$

X The equations represent a line and a circle that do not cross.

Y The equations represent a circle and a line that is a tangent to the circle.

Z The equations represent a line and a circle that intersect at two points.

7 Solve these pairs of simultaneous equations, giving your solutions
correct to two decimal places where appropriate.

(a) $x^2 + y^2 = 45$
$y - x = 3$

(b) $y = 2x + 1$
$x^2 + y^2 = 2$

(c) $x^2 + y^2 = 19$
$2x + y = 3$

8 Solve these pairs of simultaneous equations.

(a) $4x^2 + y^2 = 18$
$2x - y = 0$

(b) $3x^2 + y^2 = 4$
$y + 3x = 2$

(c) $y - x = 1$
$4y^2 - x^2 = 7$

*9 Find the exact coordinates of all the points where
the parabola $y = x^2 - 4$ intersects the circle $x^2 + y^2 = 6$.

29 Congruent triangles

A 'Fixing' a triangle

1 The triangles below are not drawn accurately.
  Find pairs of congruent triangles.
  Give the reason (SSS, …) in each case.

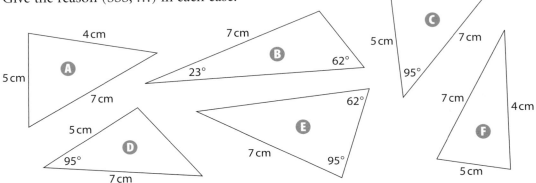

B Proving that two triangles are congruent

1 PQ and RS are line segments.
  M is the mid-point of both line segments.
  Prove that triangle PMR is congruent to triangle QMS.

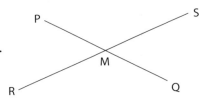

2 Points A, B, C and D lie on a circle.
  Triangle BCE is equilateral.
  Prove that triangle ABE is congruent to triangle DCE.

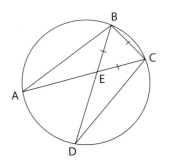

3 In this diagram, PM is perpendicular to the line *l*.
  A is a point such that PA = PM.
  The bisector of angle APM meets the line *l* at X.

  (a) Prove that triangles APX and MPX are congruent.

  (b) What can you deduce about angle PAX?

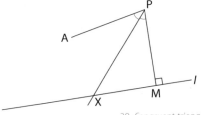

c Proving by congruent triangles

1 ABC is an isosceles triangle, with AB = AC.
BM is perpendicular to AC.
CN is parallel to BA.
AN is perpendicular to CN.

By first proving that triangles ABM and CAN are congruent, prove that BM = AN.

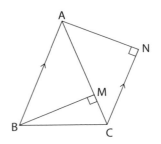

2 O is the centre of both circles.
AB is a diameter of the smaller circle.
CD is a diameter of the larger circle.

(a) Prove that angle ADO = angle BCO.

(b) What can you deduce about lines AD and CB?

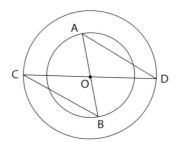

3 Line segments PQ and RS are parallel and equal.
QX and RY are perpendicular to PS.
By using congruent triangles, prove that PX = SY.

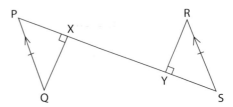

D Justifying ruler-and-compasses constructions

1 This is a construction for drawing at a point X an angle equal to a given angle BAC.

- Draw an arc of a circle with centre A to cut AB at P and AC at Q.

- Draw an arc of the same radius with centre X.

- Set the compasses to a radius equal to PQ. Choose a point R on the arc with centre X. Draw the arc RS. Then angle RXS = angle BAC.

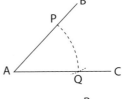

Use congruent triangles to prove that this construction is correct.

30 Proof

A Proving a statement is false with a counterexample
B Algebraic proof

1 Callum says, 'The square of any prime number is always one more than a multiple of 6.'
  Give an example to show that Callum is wrong.

2 Show that each statement below is false.

  (a) $24x \geq 24$ for all values of x.   (b) $5^x \geq 1$ for all values of x.

  (c) $\frac{1}{2}x < x$ for all values of x.   (d) $6x + 1$ is prime for all positive integers, x.

3 Show that the statement '$\dfrac{a + b}{c + d} = \dfrac{a}{c} + \dfrac{b}{d}$ for all values of a, b, c and d' is false.

4 Show that the mean of any four consecutive integers is not always an integer itself.

5 P and Q are both odd numbers.
  Which of these statements are true?

  A $P + Q$ is even.   B PQ is odd.   C $P^2 - 1$ is odd.

  D $5Q + 3$ is even.   E $P(Q - 1)$ is odd.   F $P^2 + Q$ is odd.

6 Prove that the difference between any two consecutive square numbers is always odd.

7 Prove that, if n is an integer, then $n^3 - n$ is always even.

8 Prove that, if n is an integer, then $(n + 2)^2 - (n - 2)^2$ is a multiple of 8.

9 Prove that the square of an even number is always a multiple of 4.

10 (a) Show that the sum of any four consecutive integers is always even.

   (b) Prove that the sum of any four consecutive integers is equal to the difference
       between the product of the greatest pair and the product of the smallest pair.

11 Prove that the mean of any five consecutive integers is an integer itself.

12 Prove that the sum of the squares of any five consecutive integers is a multiple of 5.

13 (a) Find the value of $n^2 + 6n$ when $n = {}^-5$.

   (b) (i) Show that $n^2 + 6n = (n + 3)^2 - 9$.

       (ii) Hence show that $n^2 + 6n \geq {}^-9$ for all values of n.

*14 Two ordinary six-sided dice are rolled.
   Prove that the difference between the product of the numbers showing on the top faces
   and the product of the numbers hidden on the bottom faces is always a multiple of 7.

31 Extending trigonometry to any triangle

This information is on the formula page that you will have in the GCSE Higher tier exam.

In any triangle ABC, area of triangle $= \frac{1}{2}ab\sin C$

sine rule: $\dfrac{a}{\sin A} = \dfrac{b}{\sin B} = \dfrac{c}{\sin C}$

cosine rule: $a^2 = b^2 + c^2 - 2bc\cos A$

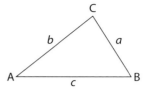

A Revision: trigonometry and Pythagoras

1 Find the length or angle marked **?**, to 1 d.p., in each of these right-angled triangles.

(a)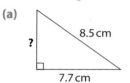

8.5 cm

?

7.7 cm

(b)

?

9.4 cm

6.8 cm

(c)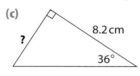

8.2 cm

?

36°

(d)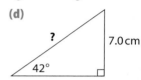

?

7.0 cm

42°

B The sine rule: finding a side
C The sine rule: finding an angle

1 Find each length marked **?** to 1 d.p.

(a)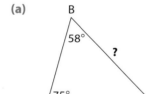

B
58°
?
A 75°
8.0 cm C

(b)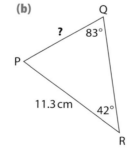

Q
? 83°
P
11.3 cm
42°
R

(c)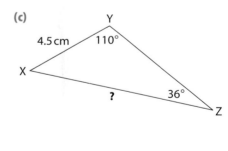

Y
4.5 cm 110°
X
? 36°
Z

2 Find the lengths marked with letters.

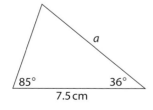

a
85° 36°
7.5 cm

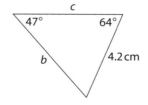

c
47° 64°
b
4.2 cm

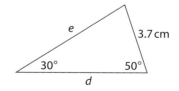

e
3.7 cm
30° 50°
d

3 Find the angle marked **?** in each triangle, to the nearest degree.
Make sure that the angle you give is possible for that triangle.
If there are two possible angles, give them both.

(a)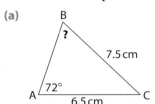

(b)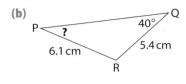

(c)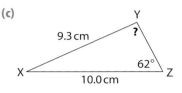

4 The diagram shows a metal framework.

(a) Find length AC.

(b) Angle ADC is obtuse.
Find the size of angle ADC.

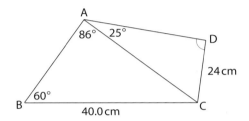

D The cosine rule

1 Find each length marked **?** to 1 d.p.

(a)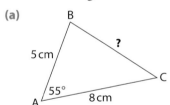

(b)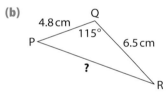

(c)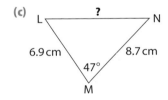

2 Find the angles marked **?**, to the nearest degree.

(a)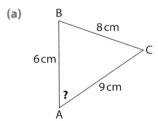

(b)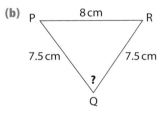

(c)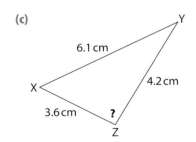

3 (a) Find length PR.

(b) Find the size of angle PSR.

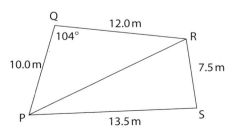

E The sine formula for the area of a triangle

1 Find the area of each of these shapes, to the nearest $0.1\,\text{cm}^2$.

(a)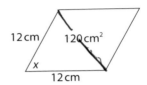

(b)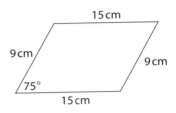

(c)

2 The area of this rhombus is $120\,\text{cm}^2$.

(a) Find the acute angle marked x.

(b) Find the length of the shorter diagonal of the rhombus.

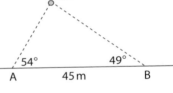

F Mixed questions

1 Two surveyors stand at points A and B, 45 m apart on a straight road.
They each record the angle between the road
and their line of sight to a tree, as shown in the diagram.

(a) Calculate the distance between surveyor A and the tree.

(b) Calculate the shortest distance from the road to the tree.

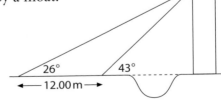

2 Town B is 10 km due north of town A.
Town C is 12 km from A and has bearing 068° from A.
Find the distance and bearing of town C from town B.

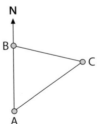

3 Find the area of this triangle.

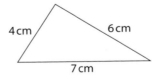

4 A tower built on horizontal ground is protected by a moat.
Roland measures the angles and distance shown.
Calculate the height of the tower.

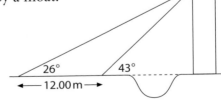

32 Transforming graphs

You need squared paper for section EF.

A Working with quadratic expressions in the form $(x + a)^2 + b$
B Completing the square

1 Find the minimum value of each expression and the value of x that gives this minimum.

(a) $(x - 5)^2 + 1$ (b) $(x + 3)^2 + 2$ (c) $(x - 1)^2 - 4$ (d) $(x + 8)^2$

2 Three of the equations below match the graphs.

W $y = (x + 2)^2 + 4$

X $y = (x - 2)^2 + 4$

Y $y = (x + 4)^2 + 2$

Z $y = (x - 4)^2 + 2$

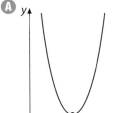
A $(4, 2)$

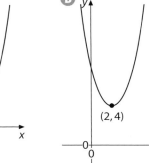
B $(2, 4)$

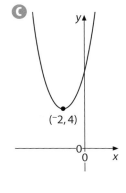
C $(-2, 4)$

(a) Match each graph with its appropriate equation.

(b) Sketch the missing graph, showing its minimum clearly.

3 Solve these equations.

(a) $(x - 5)^2 = 49$ (b) $(x + 3)^2 = 9$ (c) $(x - 1)^2 - 4 = 0$

(d) $(x + 8)^2 - 16 = 0$ (e) $(x - 4)^2 - 1 = 0$ (f) $(x + 7)^2 = 0$

4 For each of the following three equations …

P $y = (x - 2)^2 - 9$ **Q** $y = (x + 4)^2 - 1$ **R** $y = (x - 5)^2 - 25$

(a) Find the coordinates of the minimum point on its graph.

(b) Work out where the graph will cut the y-axis.

(c) Work out where the graph will cut the x-axis.

(d) Sketch the graph marking all the points you have found.

5 Write each expression in completed-square form, then work out its minimum value.

(a) $x^2 + 8x + 20$ (b) $x^2 + 4x + 5$ (c) $x^2 + 10x + 5$

(d) $x^2 + 6x - 4$ (e) $x^2 + 2x - 1$ (f) $x^2 - 4x + 9$

(g) $x^2 - 6x + 5$ (h) $x^2 - 8x + 1$ (i) $x^2 - 2x - 3$

6 Write each expression below in the form $(x + a)^2 + b$, where a and b are constants. Write down the value of a and b each time.

(a) $x^2 + 10x + 40$ (b) $x^2 + 6x + 10$ (c) $x^2 + 2x - 7$

(d) $x^2 - 4x + 13$ (e) $x^2 - 8x + 9$ (f) $x^2 - 14x - 1$

7 The expression $x^2 - 10x + 28$ can be written in the form $(x - a)^2 + b$.

(a) Find the values of a and b.

(b) Hence find the minimum point on the graph of $y = x^2 - 10x + 28$.

(c) Sketch this graph, showing the minimum point and the y-intercept.

8 The expression $x^2 - 20x$ can be written in the form $(x - p)^2 - q$. Find the values of p and q and hence the minimum value of the expression.

c Transforming quadratic graphs
D Transforming graphs in general

1 Each of these sketches shows the graph of $y = x^2$ after a translation. Find the equation of each curve.

(a)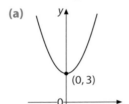

(b)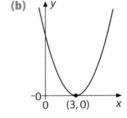

(c)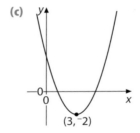

(d)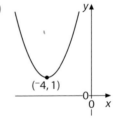

2 What is the equation of the graph of $y = x^2$ after each of these transformations?

(a) A translation of 9 units to the left

(b) Reflection in the x-axis

(c) A translation of $\begin{bmatrix} -2 \\ -5 \end{bmatrix}$

(d) A stretch by a factor of 5 in the y-direction

(e) A stretch by a factor of 5 in the x-direction

3 Sketch the image of $y = x^2$ after a reflection in the x-axis followed by a translation of 10 units vertically. Label the transformed curve with its equation.

4 Sketch the image of $y = x^3$ after a reflection in the x-axis followed by a horizontal translation of 1 unit to the right. Label the transformed curve with its equation.

5 Write down the equation of $y = \frac{1}{x}$ after a stretch by a factor of 4 in the y-direction.

6 The graph of $y = \sin x$ is shown for $^{-}180° \le x \le 180°$.
On a separate diagram for each pair, sketch
these graphs for $^{-}180° \le x \le 180°$.

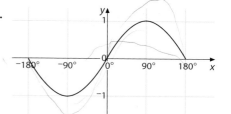

(a) $y = 2\sin x$ and $y = ^{-}2\sin x$

(b) $y = \sin\frac{1}{2}x$ and $y = ^{-}\sin\frac{1}{2}x$

(c) $y = ^{-}\sin x$ and $y = 2 - \sin x$

7 Sketch the graph of $y = \sin(x - 90°)$.

8 Each graph is the image of $y = \cos x$ after a transformation.
Write the equation of each graph.

(a)
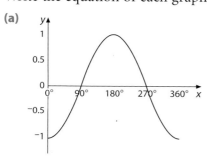

(b)
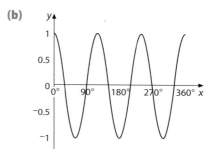

E Function notation
F Transforming functions

1 Given that $f(x) = 4x + 1$, evaluate   (a) $f(2)$ (b) $f(^{-}1)$ (c) $f\left(^{-}\frac{1}{4}\right)$

2 Given that $g(x) = x^2 - 2$, evaluate   (a) $g(3)$ (b) $g(0)$ (c) $g(^{-}2)$

3 (a) Given that $f(x) = x^3$, evaluate

     (i) $f(2)$ (ii) $f(0)$ (iii) $f(^{-}3)$

(b) Given that $y = f(x + 2)$, find the value of y when

     (i) $x = 0$ (ii) $x = ^{-}5$ (iii) $x = 2$

(c) Which of these gives the rule for y in terms of x?

     A $y = x^3 + 2$ B $y = x^3 - 2$ C $y = (x + 2)^3$ D $y = (x - 2)^3$

4 $f(x) = x^2$ and $y = f(2x)$.

(a) Find the value of y when   (i) $x = 1$ (ii) $x = 2$ (iii) $x = ^{-}5$

(b) Write the rule for y in terms of x.

5 $f(x) = \cos x$ and $y = f(x - 90°)$

(a) Write the rule for y in terms of x.

(b) Sketch the graph of $y = f(x - 90°)$ for $0° \le x \le 360°$.

6 Given that $f(x) = x^2 - 3$, write each of these rules for y in terms of x.
Expand any brackets and write each rule in its simplest form.

(a) $y = f(x) + 2$ (b) $y = f(x + 2)$ (c) $y = 2f(x)$ (d) $y = f(2x)$

7 The diagram shows the graph of $y = f(x)$.
On separate grids draw the graph of
each of the transformed functions below.

In each case label the transformed point P′
with its coordinates.

(a) $y = f(x) + 2$ (b) $y = f(x + 2)$

(c) $y = {}^-f(x)$ (d) $y = \frac{1}{2}f(x)$

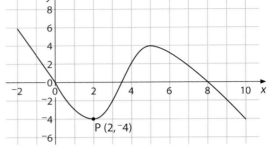

8 The diagram shows the graph of $y = f(x)$.
The point marked A is the maximum point on the graph.

Write down the coordinates for the maximum point for
each of these curves.

(a) $y = f(x) - 1$ (b) $y = f(x - 1)$

(c) $y = 2f(x)$ (d) $y = f(2x)$

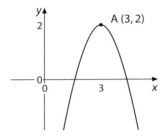

9 The diagram shows the graph of $y = f(x)$.

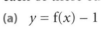

(a) On separate grids draw each of these for ${}^-4 \le x \le 12$.

(i) $y = \frac{1}{2}f(x)$ (ii) $y = f(x) + 1$ (iii) $y = f\left(\frac{1}{2}x\right)$ (iv) $y = 2 - f(x)$

(b) The two graphs below are transformations of $y = f(x)$.
Choose the correct equation for each graph.

A $y = \frac{1}{4}f(x)$ B $y = 4f(x)$ C $y = f(x + 4)$ D $y = f(x - 4)$

E $y = f\left(\frac{1}{4}x\right)$ F $y = f(4x)$ G $y = f(x) + 4$ H $y = f(x) - 4$

(i) 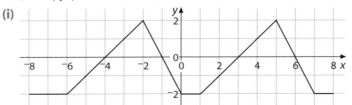 (ii)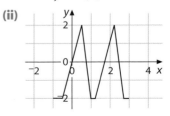

33 Rational and irrational numbers

A Review: fractions to decimals
B Recurring decimals to fractions

1 Write these fractions as decimals.

(a) $\frac{12}{25}$ (b) $\frac{17}{20}$ (c) $\frac{7}{8}$ (d) $\frac{13}{40}$ (e) $\frac{9}{80}$

2 Which of these fractions are equivalent to recurring decimals?

$\frac{7}{30}$ $\frac{5}{32}$ $\frac{13}{90}$ $\frac{8}{27}$ $\frac{9}{40}$ $\frac{1}{45}$ $\frac{11}{75}$ $\frac{5}{7}$ $\frac{17}{80}$

3 Write these fractions as decimals.

(a) $\frac{1}{6}$ (b) $\frac{2}{9}$ (c) $\frac{4}{11}$ (d) $\frac{3}{7}$ (e) $\frac{7}{60}$

4 Write these decimals as fractions in their simplest form.

(a) 0.4   (b) 0.65   (c) 0.08   (d) 0.135   (e) 0.084

5 Write these as fractions in their lowest terms.

(a) $0.\dot{4}$ (b) $0.454\,545\ldots$ (c) $0.\dot{1}\dot{2}$ (d) $0.162\,162\ldots$

(e) $0.388\,888\ldots$ (f) $0.1\dot{5}$ (g) $0.0\dot{5}\dot{2}$ (h) $0.4\dot{5}4\dot{3}$

6 (a) Write $0.\dot{6}$ as a fraction.   (b) Hence write $0.0\dot{6}$ as a fraction.

7 (a) Write $0.\dot{7}$ as a fraction.   (b) Write 0.3 as a fraction.

(c) Hence write $0.4\dot{7}$ as a fraction.

*8 The sum of the reciprocals of two consecutive numbers is $0.1\dot{9}\dot{0}$. Find the numbers.

C Irrational numbers

1 Which of these numbers are irrational?

$1.\dot{3}$ $\sqrt{10}$ $\frac{22}{7}$ 2π $\sqrt{49}$ $\sqrt[3]{27}$ 4.9317 $\sqrt[3]{9}$

2 Simplify these.

(a) $\sqrt{3} + \sqrt{3}$ (b) $4(\sqrt{3} + 5)$ (c) $7\sqrt{5} - 3\sqrt{5}$

(d) $(\sqrt{6})^2$ (e) $(4\sqrt{3} + 5) - (\sqrt{3} - 4)$ (f) $(2\sqrt{5} + \sqrt{6}) + (3\sqrt{5} - 2\sqrt{6})$

3 Write down an irrational number between 6 and 7.

4 (a) For each of these decimals, state with reasons whether it is a rational or an irrational number.

(i) 0.8   (ii) 0.828 228 222 822 22…   (iii) $0.\dot{8}2\dot{8}$

(b) Write, where possible, the decimals in (a) as fractions in their simplest form.

5 Write down the exact length of the sides of a square with area $20\,\text{cm}^2$.

6 What is the exact length of side x of this triangle?

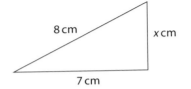

8 cm   x cm   7 cm

1 Simplify these.

(a) $\sqrt{5} \times \sqrt{7}$ (b) $2\sqrt{5} \times \sqrt{5}$ (c) $\sqrt{2} \times \sqrt{8}$ (d) $3\sqrt{5} \times 2\sqrt{6}$

2 Write each of these in the form $p\sqrt{q}$, where q is the smallest integer possible.

(a) $\sqrt{20}$ (b) $\sqrt{75}$ (c) $\sqrt{90}$ (d) $\sqrt{54}$ (e) $\sqrt{250}$

(f) $2\sqrt{24}$ (g) $5\sqrt{8}$ (h) $2\sqrt{40}$ (i) $5\sqrt{128}$ (j) $5\sqrt{700}$

3 Write each of these in the form $p\sqrt{q}$, where q is the smallest integer possible.

(a) $\sqrt{3} \times \sqrt{6}$ (b) $\sqrt{5} \times \sqrt{35}$ (c) $\sqrt{20} \times \sqrt{10}$ (d) $\sqrt{21} \times \sqrt{7}$

4 Simplify these. $2\sqrt{3} + \sqrt{3} = 3\sqrt{3}$

(a) $\sqrt{3} + \sqrt{12}$ (b) $\sqrt{125} - 2\sqrt{5}$ (c) $\sqrt{60} + \sqrt{15}$ (d) $2\sqrt{80} - \sqrt{45}$

5 (a) Write $\sqrt{32}$ in the form $a\sqrt{b}$, where b is a prime number.

(b) Hence simplify $\dfrac{\sqrt{32}}{2}$.

6 Simplify these.

(a) $\dfrac{\sqrt{40}}{2}$ (b) $\dfrac{\sqrt{50}}{\sqrt{2}}$ (c) $\dfrac{6}{\sqrt{6}}$ (d) $\dfrac{\sqrt{150}}{5}$ (e) $\dfrac{\sqrt{48}}{2}$

7 Multiply out the brackets and simplify these.

(a) $\left(4 + \sqrt{5}\right)\left(4 - \sqrt{5}\right)$ (b) $\left(6 - \sqrt{2}\right)\left(1 + \sqrt{2}\right)$ (c) $\left(5 + \sqrt{3}\right)^2$

8 What is the area of a square of side $\sqrt{3} - 1$?
Give your answer in the form $a - b\sqrt{3}$.

9 (a) Simplify these.

(i) $(\sqrt{7} + \sqrt{2})(\sqrt{7} - \sqrt{2})$ (ii) $(\sqrt{5} + \sqrt{20})^2$ (iii) $(\sqrt{8} - \sqrt{2})(\sqrt{12} + \sqrt{2})$

(b) Which of your answers to (a) are irrational?

10 Write $(\sqrt{10} + \sqrt{2})^2$ in the form $a + b\sqrt{c}$, where c is as small as possible.

11 (a) Given that $x = \sqrt{10} - 2$, find the value of x^2 in the form $a + b\sqrt{c}$.

(b) Hence, or otherwise, show that $x = \sqrt{10} - 2$ is a solution to
the equation $x^2 + 4x - 6 = 0$.

12 Show that $(\sqrt{12} + \sqrt{3})^2$ is rational.

F Further simplifying with square roots

1 Simplify   (a) $\dfrac{\sqrt{20}}{\sqrt{5}}$ (b) $\dfrac{\sqrt{32}}{\sqrt{2}}$ (c) $\dfrac{\sqrt{21}}{\sqrt{3}}$

2 Evaluate   (a) $\sqrt{\dfrac{16}{49}}$ (b) $\sqrt{\dfrac{1}{25}}$ (c) $\sqrt{\dfrac{64}{9}}$

3 Write $\sqrt{\dfrac{7}{25}}$ in the form $\dfrac{\sqrt{a}}{b}$ where a and b are integers.

4 For each expression, rationalise the denominator and write the result
in its simplest form.

(a) $\dfrac{2}{\sqrt{3}}$ (b) $\dfrac{1}{\sqrt{5}}$ (c) $\dfrac{11}{\sqrt{11}}$ (d) $\dfrac{\sqrt{7}}{\sqrt{2}}$ (e) $\dfrac{\sqrt{6}}{\sqrt{5}}$

5 Rationalise each denominator.

(a) $\dfrac{\sqrt{5} + 1}{\sqrt{5}}$ (b) $\dfrac{8 - \sqrt{2}}{\sqrt{2}}$ (c) $\dfrac{\sqrt{7} - 4}{\sqrt{7}}$

6 Show that $\dfrac{\sqrt{3}}{5} + \dfrac{1}{\sqrt{3}} = \dfrac{8\sqrt{3}}{15}$.

G Surd solutions to quadratic equations

1 Solve each of these equations, giving the solution in the form $x = p \pm \sqrt{q}$.

(a) $x^2 - 4x - 1 = 0$ (b) $x^2 - 2x - 6 = 0$ (c) $x^2 + 10x + 19 = 0$

2 (a) Write the expression $x^2 - 6x + 1$ in the form $(x - p)^2 - q$.

(b) Hence solve the equation $x^2 - 6x + 1 = 0$, giving the solution in
the form $x = a \pm b\sqrt{c}$ where c is as small as possible.

3 Solve each of these equations by completing the square, giving the solution in surd form.

(a) $x^2 + 8x + 9 = 0$ (b) $x^2 - 6x - 11 = 0$ (c) $x^2 + 12x + 4 = 0$

4 Use the quadratic formula to solve each of these equations, giving the solution in the form $x = a \pm b\sqrt{c}$.

(a) $2x^2 - 8x - 1 = 0$ (b) $2x^2 - 5x + 1 = 0$ (c) $3x^2 + 4x - 2 = 0$

H Mixed questions

1 A large square encloses two smaller squares as shown.
The smaller squares have areas $3\,\text{cm}^2$ and $5\,\text{cm}^2$.

(a) Write down the exact length of a side of the largest square.

(b) Work out the area of the largest square, writing your answer in the form $a \pm b\sqrt{c}$.

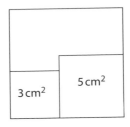

2 Which of the following are irrational?

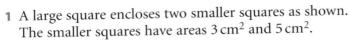

$(\sqrt{7})^2$ $\sqrt{5} - 2$ $\sqrt{28} \times \sqrt{14}$ $(\sqrt{5} - 2)(\sqrt{5} + 2)$ $\dfrac{\sqrt{12}}{\sqrt{3}}$ $\sqrt{24} - \sqrt{6}$

3 (a) Find, in surd form, the length labelled a.

(b) Do the same for b and explain why $b = 2a$.

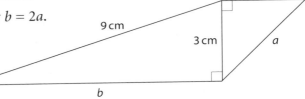

4 (a) Evaluate   (i) $\sqrt{6\frac{1}{4}}$ (ii) $\sqrt{0.01}$

(b) Write $\sqrt{2\frac{2}{9}}$ in the form $\dfrac{a\sqrt{b}}{c}$ where a, b and c are integers.

5 The diagram shows a square with the quadrant of a circle shaded.
The area of the square is $6\,\text{cm}^2$.
What is the exact value of the shaded area?

6 Rita thinks that if two irrational numbers are added the answer will always be irrational. Give an example to show when this is not true.

7 Show that $\dfrac{\sqrt{27} + \sqrt{12}}{\sqrt{300} - \sqrt{75}} = 1$.

34 Pythagoras and trigonometry in three dimensions

A Solving problems using a grid

1 Each grid here consists of unit squares.
Calculate the following, giving your answers to 2 d.p.

(a) AB

(b) AC

(c) The angle between line AB and line AC

(d) The angle between line AC and plane OBCE

(e) AD

(f) DC

(g) Angle CAD

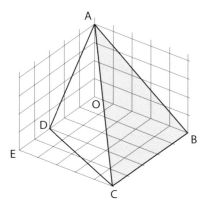

B Solving problems without a grid

1 A pole of height 5.2 m is kept vertical by three guy ropes.
Each guy rope is fixed to the top of the pole and to a point
on level ground 2.4 m from the base of the pole.
The three points where the ropes are fixed to the ground
form an equilateral triangle.

Calculate these, correct to 1 d.p.

(a) The length of each rope

(b) The angle each rope makes with the ground

(c) The length of each side of the equilateral triangle

(d) The angle between any pair of ropes

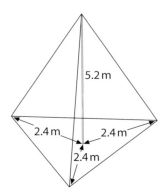

2 The diagram shows a cuboid.

(a) Find the length of the diagonal AG.

(b) Find the angle between AG and the plane ADHE.

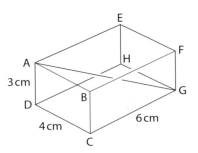

3 A square-based right pyramid has edges with lengths as shown.

(a) Find the height of the pyramid.

(b) Calculate the volume of the pyramid.

(c) Calculate the angle between each sloping face and the base.

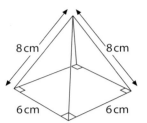

*4 The diagram shows the roof of a house.
The base is a rectangle.
The sloping ends are isosceles triangles.
Each side of the roof is an isosceles trapezium.

(a) Find the total surface area of the roof.

(b) Find the volume enclosed by the roof.

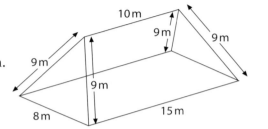

---

c Using coordinates in three dimensions

1 The point P (3, 2, 6) is shown.

(a) What is the length of the line segment from the origin O (0, 0, 0) to point P?

(b) Give the mid-point of the line segment from O to P.

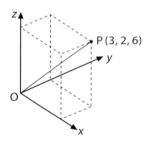

2 Find **(i)** the length (to 2 d.p. if necessary) and **(ii)** the mid-point of these line segments.

(a) From (0, 0, 0) to (16, 2, 8)   (b) From (0, 0, 0) to (6, 8, 3)

3 The points C (4, 2, 1) and D (6, 8, 10) are shown.

(a) Find the length of the line segment from C to D.

(b) Give the mid-point of the line segment from C to D.

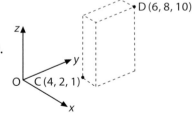

4 Find **(i)** the length (to 2 d.p. if necessary) and **(ii)** the mid-point of these line segments.

(a) From (5, 8, 0) to (9, 1, 4)   (b) From (6, 6, 3) to (10, 4, 8)

5 Look back at the diagram for question 1.
Find the angle between the line segment OP and the plane in which the x- and y-axes lie.

6 Three points are given as E (3, 1, 0), F (4, 5, 9) and G (7, 10, 1).
Find the lengths EF, FG and GE as exact values. What type of triangle is EFG?

Mixed practice 5

You need a pair of compasses, squared paper and graph paper.

1 In this diagram PQ is a diameter of the circle.

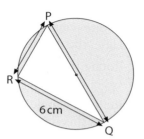

   (a) Show that PRQ is a right-angled triangle.

Tan $\angle$ QPR = 1.5

   (b) Find the length PR.

   (c) Show that the radius of the circle is $\sqrt{13}$.

   (d) Find the shaded area in cm^2 as an exact value.

2 (a) Write down two consecutive numbers and add them together.

   (b) Square both consecutive numbers and find the difference between the squares.

   (c) What do you notice about the answers to (a) and (b)?
Prove that this will happen with any two consecutive numbers.

3 Do not rub out your construction lines in this question.

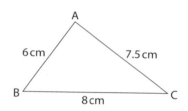

   (a) Use a ruler and compasses to construct this triangle.

   (b) The point P is equidistant from AB and AC.
It is also equidistant from A and B.
Use a straight edge and compasses to find
the position of the point P.

   (c) Use trigonometry to calculate the size of the largest angle in your triangle.

4 Work out each of these, giving your answer in standard form.

   (a) $(4 \times 10^5) \times (5 \times 10^6)$ **(b)** $(3 \times 10^{-4})^2$ **(c)** $\dfrac{4 \times 10^7}{8 \times 10^5}$

5 (a) (i) Expand and simplify $(n + 7)(n + 3)$.

     (ii) Hence find the value of 37×33 without using a calculator.

   (b) Expand and simplify $(x + 6)^2$ and use the result to find the value of 26^2.

6 A museum has a collection of 34 old cars. Some have three wheels and some four.
The number of three-wheeled cars is n.
The total number of wheels on all the cars is 125.

   (a) Write an expression in terms of n for the number of four-wheeled cars.

   (b) Use the information about the total number of wheels to form an equation
and solve it to find n.

7 Write $(3 + \sqrt{5})(2 + \sqrt{20})$ in the form $p + q\sqrt{r}$, where p, q and r are positive integers
and r is as small as possible.

8 If two integers differ by 2, prove algebraically that their squares differ by a multiple of 4.

9 A piece of zinc with a volume of $0.54 \, \text{m}^3$ has a mass of $3856 \, \text{kg}$.
Calculate the density of zinc, in g/cm^3, correct to one decimal place.

10 Copy the grid and triangles A and B.

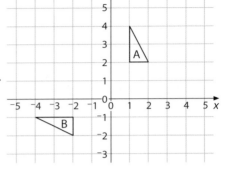

   (a) Describe the transformation that maps A to B.

   (b) Rotate triangle A 90° clockwise about $(1, 1)$.
       Label the image C.

   (c) Rotate triangle C 90° anticlockwise about $(0, {}^-1)$.
       Label this image D.

   (d) Reflect triangle D in the line $y = x$.
       Label this image E.

   (e) Describe fully the single transformation which
       maps B to E.

11 This table shows the distribution of weights of
a group of babies at their 6-month health check.

   (a) Draw a frequency polygon for this data.

   (b) Estimate the mean weight of these babies.

   (c) Make a cumulative frequency table.

   (d) Draw a cumulative frequency graph
       of the distribution.

   (e) Use your graph to estimate

      (i) the median weight

      (ii) the interquartile range

Weight (w kg)	Frequency
$6.0 < w \leq 6.5$	8
$6.5 < w \leq 7.0$	10
$7.0 < w \leq 7.5$	24
$7.5 < w \leq 8.0$	35
$8.0 < w \leq 8.5$	33
$8.5 < w \leq 9.0$	21
$9.0 < w \leq 9.5$	15
$9.5 < w \leq 10.0$	4

12 The function $f(x)$ is defined for values of x in
the interval $^-3 \leq x \leq 5$.
The graph of $y = f(x)$ is shown in this diagram.

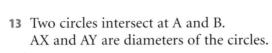

Draw a graph for each of these.

   (a) $y = {}^-f(x)$ **(b)** $y = f({}^-x)$

   (c) $y = f(x) + 1$ **(d)** $y = f(x + 1)$

   (e) $y = \frac{1}{2}f(x)$ **(f)** $y = f\left(\frac{1}{2}x\right)$

13 Two circles intersect at A and B.
AX and AY are diameters of the circles.

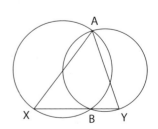

   (a) Prove that XBY is a straight line.

   (b) Given that $AX = 8$, $AY = \sqrt{63}$ and $AB = 6$, calculate the
       area of triangle AXY as an exact value in its simplest form.

14 Sketch the circle that has equation $x^2 + y^2 = 9$.

15 The formula for the nth triangle number is $\frac{1}{2}n(n+1)$.

   (a) What is the sum of the 9th and 10th triangle numbers?

   (b) Write down an expression for the $(n+1)$th triangle number.

   (c) Prove that adding two consecutive triangle numbers gives a square number.

16 In the diagram AD is a tangent to the circle.

   (a) Prove that triangles ABD and CAD are similar.

   (b) Calculate the exact length of AD.

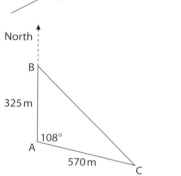

17 A, B and C are three points on level ground.
B is 325 m due north of A.
C is 570 m from A on a bearing of 108° from A.

  Calculate these.

   (a) Area ABC, to the nearest $10 \, \text{m}^2$

   (b) Distance BC, to the nearest metre

   (c) The bearing of C from B, to the nearest 0.1°

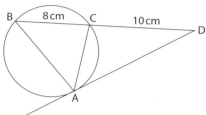

18 Write each of these as a fraction in its simplest form.

   (a) $0.5\dot{4}$ **(b)** $0.\dot{3}2\dot{1}$ **(c)** $0.4\dot{8}$

19 The line segment AB has end-points A $(1, 4, 3)$ and B $(5, 0, {}^-3)$.

   (a) Find the coordinates of the mid-point of AB.

   (b) Find the exact value of the length of the line segment AB.

20 A square-based pyramid has a base 8 cm square and a height of 8 cm.

  Calculate, to two significant figures,

   (a) the length of each sloping edge

   (b) the area of each sloping face

   (c) the angle between each sloping face and the base

   (d) the angle between each sloping edge and the base

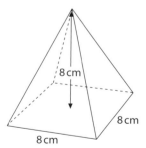

21 (a) Find the coordinates of the points of intersection of the line $y + 2x = 5$ and the circle $x^2 + y^2 = 50$.

   (b) Find the length of the chord joining these two points of intersection, giving your answer in the form $a\sqrt{b}$, where a and b are integers with b as small as possible.

22 Rationalise the denominator of $\dfrac{\sqrt{5}+4}{\sqrt{5}}$. Simplify your answer fully.

23 (a) Rearrange the formula $\dfrac{1}{a} + \dfrac{1}{b} = \dfrac{1}{h}$ to make h the subject.

(b) Calculate h given that $a = 5$ and $b = 3$.

24 (a) Show that $x = 4$ is a solution of the equation $x^2 = 2^x$.

(b) (i) On the same set of axes, draw graphs of $y = x^2$ and $y = 2^x$ for $^-3 \leq x \leq 3$.

(ii) Use your graphs to show that there is a solution to $x^2 = 2^x$ between $^-1$ and 0.

(iii) Use trial and improvement to find this solution, correct to 1 d.p.

25 Prove that the line $2x + y = 10$ does not intersect the circle $x^2 + y^2 = 15$.

26

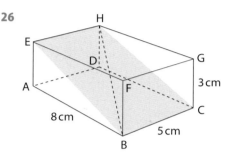

The diagram shows a cuboid ABCDEFGH whose edges are of length 8 cm, 5 cm and 3 cm.

Calculate the following.

(a) The angle between the plane BCHE and the base ABCD

(b) The length of the diagonal BH

(c) The angle between BH and the plane CDHG

27 (a) The expression $2x^2 - 20x + 53$ can be written in the form $2(x - a)^2 + b$. Find the values of a and b.

(b) Sketch the graph of $y = 2x^2 - 20x + 53$, showing the minimum point and y-intercept.

28 In this diagram a circle with centre O touches all three sides of an isosceles triangle ABC. The circle touches the base at M and touches the other two sides at N and P. Show that ONB and OMC are congruent triangles.

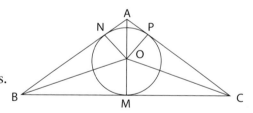

*29 (a) Find the area of the shaded square.

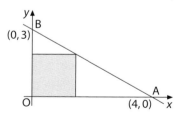

(b) AB touches the circle at T. OT is a diameter of the circle. Find, in terms of π, the area of the circle.

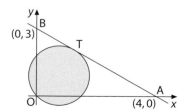